I0427221

Fundamentals of Engineering Mechanics

3rd Edition

Basic Concepts in:
Statics

David A. Cicci
Darby A. Cicci

Copyright © 2024 by David A. Cicci and Darby A. Cicci

ISBN 9798883759641

All rights reserved. No part of this publication may be reproduced, distributed, or transmitted in any form or by any means, including photocopying, recording, or other electronic or mechanical methods, without the prior written permission of the publisher at the address below. Permission requests should be addressed to the publisher at the email address below. Violations of the copyright laws of the United States by any of the actions stated above will be prosecuted to the fullest extent of the law.

Dynamic Solutions Publishing
Auburn, AL
ciccida@auburn.edu

Printed in the United States of America

PREFACE TO THE 3RD EDITION

The 1st Edition of *Fundamentals of Engineering Mechanics, Basic Concepts in: Statics, Mechanics of Materials, and Dynamics* was formally published in 2019, although its origin dates back to 2001 when it was published internally for use by students in the College of Engineering at Auburn University. The 2nd Edition was published in 2022, which provided a series of revisions and improvements to the 1st Edition.

Since the 1st Edition was published containing the three subjects together, five additional volumes were created covering one or two of the subjects rather than all three subjects together in one volume. This was done to provide a subset of the material to individuals who didn't need information on all three subjects or for use in courses covering one or two subjects. That is, separate volumes were created under the *Fundamentals of Engineering Mechanics* title, with subtitles of: *Basic Concepts in: Statics; Basic Concepts in: Mechanics of Materials; Basic Concepts in: Dynamics; Basic Concepts in: Statics and Dynamics;* and *Basic Concepts in: Statics and Mechanics of Materials*. These offerings were consistent with the authors' original intent to provide quality information in textbook form at minimal cost to students.

The 2nd Edition of each volume included eight problems with answers at the end of each module. Based on feedback received, the authors decided to publish a 3rd Edition to add include additional problems. This new editions added seven problems at the end of each module, bringing the number of problems included to fifteen per module in the 3rd Edition. of each volume.

Other than the additional problems included, the information presented in each module is largely unchanged from the previous edition and still includes several problems with solutions, based on the concepts discussed.

David A. Cicci, Ph.D.
Professor Emeritus
Department of Aerospace Engineering
Auburn University
Auburn, AL

(This page was intentionally left blank.)

PREFACE TO THE 2ND EDITION

The 1st Edition of *Fundamentals of Engineering Mechanics, Basic Concepts in: Statics, Mechanics of Materials, and Dynamics* was formally published in 2019, although its origin dates back to 2001 when it was published internally for use by students in the College of Engineering at Auburn University. That being the case, the authors felt it was time for a 2nd Edition to provide a series of revisions and improvements to the original work.

This 2nd Edition utilizes an updated equation editor to convert the previous in-line equations to professional-type equations for better readability and to facilitate better understanding. In doing so, fonts were changed throughout the text to provide consistency with the equations. Some of the figures presented were also refreshed for simplicity and clarification. The overall book design and formatting were updated for improved aesthetics as well.

In addition, many of the problems contained in each module were modified, enhanced, and grouped with problems using the same or similar figures to minimize redundancy. Clarifications were also provided throughout and a few of the answers to problems which were found to be inaccurate were corrected.

Since the 1st Edition was published containing the three subjects together, five additional volumes were created covering one or two of the subjects rather than all three subjects together in one volume. This was done to provide a subset of the material to individuals who didn't need information on all three subjects or for use in courses covering one or two subjects. That is, separate volumes were created under the *Fundamentals of Engineering Mechanics* title, with subtitles of: *Basic Concepts in: Statics; Basic Concepts in: Mechanics of Materials; Basic Concepts in: Dynamics; Basic Concepts in: Statics and Dynamics;* and *Basic Concepts in: Statics and Mechanics of Materials*. These offerings are consistent with the authors' original intent to provide quality information in textbook form at minimal cost to students. 2nd Editions of each of those volumes are now available.

(This page was intentionally left blank.)

PREFACE TO THE 1ST EDITION

Fundamentals of Engineering Mechanics, Basic Concepts in Statics, Mechanics of Materials, and Dynamics was developed to present three topics in engineering mechanics: statics, mechanics of materials, and dynamics, to undergraduate students in non-mechanics oriented engineering disciplines. It was not originally designed for students in the more mechanics-based engineering disciplines such as aerospace, mechanical, or civil engineering.

The three subjects presented are traditionally taught in a three-course sequence of undergraduate courses which allows for in-depth study and a more comprehensive learning experience. The result of combining three subjects into a one semester-long course is that a very limited amount of time is available for each subject. Therefore, a less in-depth approach to the subject matter is presented in this textbook. It's primarily intended as an introduction for individuals with little or no background in statics, mechanics of materials, and dynamics, although it is assumed that students have a strong background in mathematics, through differential and integral calculus and differential equations. In addition, this textbook can also be used as a review for students or professionals who've previously been exposed to these subjects. That being the case, it can be used as a preparation for the Fundamentals of Engineering (FE) Examination or the Professional Engineers (PE) Examination, both of which are required for board certification of practicing engineers.

This publication presents the subject matter in a module-based learning approach. Each module presents a few concepts in the manner that an instructor might present the material during in-class lectures. The basic concepts are explained using simple illustrations to help facilitate understanding of the subjects. Example problems are included within the module to demonstrate the concepts and enhance the learning experience. A problem set, with answers provided, is also included at the end of each module. The material is presented from an applications-oriented viewpoint with a minimal amount of accompanying theory and mathematical derivations of the formulas. This type of presentation is more suitable for the students to develop the understanding of the applications encountered on the FE and PE examinations.

If this textbook is used for an academic course, the material can be covered in a full semester consisting of approximately 15 weeks, or 45 class-hours, and each module can generally be covered in a one-hour period. The three subjects presented are broken down as follows: 13 modules in statics, 13 modules in mechanics of materials, and 14 modules in dynamics. The additional 5 class-hours could be used for examinations/quizzes, reviews, more in-depth discussions of certain material, or the presentation of additional material of interest to the students.

The instructor may also choose to provide additional lecture material to cover certain modules in more detail or skip some modules to modify the course content as desired. Flexibility is built into the module-based structure to allow the instructor freedom to adjust the syllabus, the number of examinations/quizzes, review sessions, and whether to include in-class exercises as part of the learning experience.

The material is presented in a way that students should be able to read and understand the concepts on their own outside of class and only meet with the class on a limited basis for questions, discussions, reviews, quizzes, or examinations, if desired by the instructor. Again, the manner in which the course is conducted is at the discretion of the instructor since the module-based approach offers flexibility in possible methods employed to adequately cover the material.

TABLE OF CONTENTS

PART I – STATICS

(This page was intentionally left blank.)

PART I - STATICS

(This page was intentionally left blank.)

MODULE 1: Introduction to Vectors

The study of engineering mechanics deals primarily with two types of quantities, scalars and vectors. Scalar quantities are ones that only have a magnitude, while vector quantities are those that have both a magnitude and a direction. Examples of scalar quantities are time, distance, area, volume, density, speed, mass, and energy. Examples of vector quantities are displacement, velocity, acceleration, force, moment, and momentum.

A vector quantity is generally indicated by a boldface character or by a lightface character having an arrow or line drawn above it, while a scalar quantity is simply indicated by a lightface character. Graphically, a vector quantity **P** is represented by a line segment in a specific direction, with an arrowhead on one end to show the sense of the vector, as provided in Figure 1.1. The magnitude of **P**, represented by |**P**| or P, is a scalar quantity.

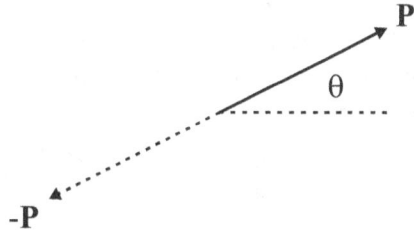

Figure 1.1

Vectors are combined using the parallelogram law. For example, the sum or resultant, **R**, of two vectors, **P**$_1$ and **P**$_2$, can be expressed analytically by the vector equation

$$\mathbf{R} = \mathbf{P}_1 + \mathbf{P}_2 = \mathbf{P}_2 + \mathbf{P}_1$$

Figure 1.2 shows this vector addition using the parallelogram law, which is obtained by placing the tail of one vector at the head of the second vector. The sum of more than two vectors follows the same procedure and the resultant will form a closed polygon having the number of sides equal to the number of forces plus one for the resultant.

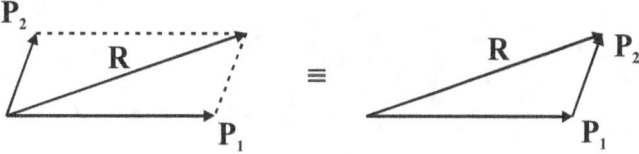

Figure 1.2

The difference between two vectors can be expressed as the sum of one positive vector and one negative vector, as shown by

$$\mathbf{R} = \mathbf{P}_1 - \mathbf{P}_2 = \mathbf{P}_1 + (-\mathbf{P}_2)$$

Graphically, the difference between vectors $\mathbf{P}_1$ and $\mathbf{P}_2$ is shown in Figure 1.3.

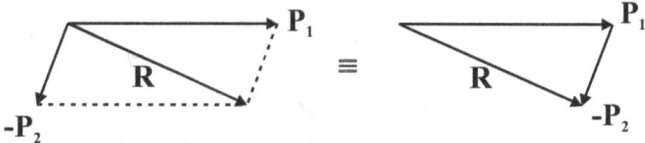

Figure 1.3

Vector quantities must be defined in a particular coordinate system. The most common is the Cartesian coordinate system, having unit vectors $\mathbf{i}$, $\mathbf{j}$, and $\mathbf{k}$ in the x-, y-, and z-directions, respectively. The magnitude of each of these unit vectors is 1. This coordinate system and the unit vectors are shown in Figure 1.4.

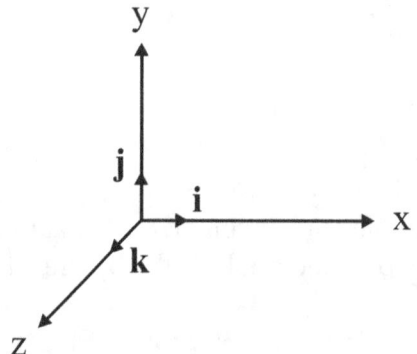

Figure 1.4

The three-dimensional vector $\mathbf{P}$ in Cartesian coordinates is shown in Figure 1.5.

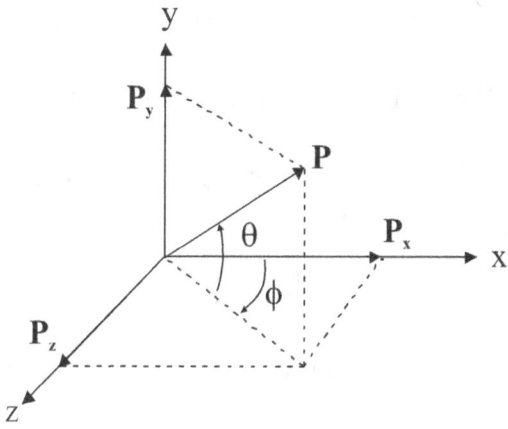

Figure 1.5

This vector can be expressed using unit vectors as

$$\mathbf{P} = P_x \mathbf{i} + P_y \mathbf{j} + P_z \mathbf{k}$$

The vector components of **P**, i.e., P_x, P_y, and P_z, are each expressed as a scalar quantity times a unit vector. The magnitudes of the vector components written in terms of angles θ and ϕ are

$$P_x = P \cos \theta \cos \phi$$

$$P_y = P \sin\theta$$

$$P_z = P \cos \theta \sin \phi$$

$$P = |\mathbf{P}| = \sqrt{(P_x)^2 + (P_x)^2 + (P_x)^2}$$

Consider the two-dimensional case shown in Figure 1.6.

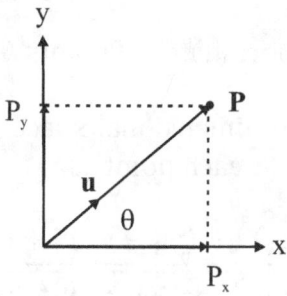

Figure 1.6

The vector **P** can be written in terms of unit vectors **i** and **j** and angle θ as

$$\mathbf{P} = P_x\,\mathbf{i} + P_y\,\mathbf{j} = P\cos\theta\,\mathbf{i} + P\sin\theta\,\mathbf{j}$$

The relationship between the vector components and the orientation angle θ is

$$\tan\theta = \frac{P_y}{P_x}$$

Vector **P** can also be written in terms of a unit vector defined in the **P** direction, **u**, as

$$\mathbf{P} = P\,\mathbf{u}$$

This unit vector can be determined from the coordinates of any two points, A and B, which lie on the line-of-action of **P**, as shown in Figure 1.7.

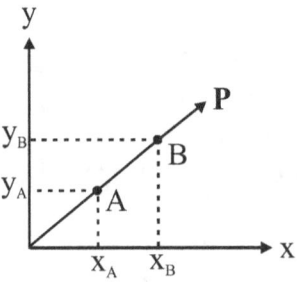

Figure 1.7

This unit vector **u** is equivalent to the vector from A to B divided by distance between points A and B and can be calculated using the formula

$$\mathbf{u} = \frac{(x_B - x_A)\,\mathbf{i} + (y_B - y_A)\,\mathbf{j}}{\sqrt{(x_B - x_A)^2 + (y_B - y_A)^2}}$$

where (x_A, y_A) and (x_B, y_B) are the coordinates of points A and B, respectively.

To determine a unit vector in three-dimensional space, the formula above is modified to include the z-direction coordinates for each point as

$$\mathbf{u} = \frac{(x_B - x_A)\,\mathbf{i} + (y_B - y_A)\,\mathbf{j} + (z_B - z_A)\,\mathbf{k}}{\sqrt{(x_B - x_A)^2 + (y_B - y_A)^2 + (z_B - z_A)^2}}$$

Example 1.1

In the figure below, the resultant of the three forces shown is 25 **j** lb. Determine the magnitudes of both **P** and **F**.

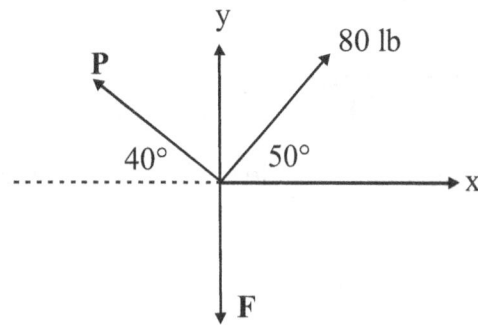

Solution:

$$\mathbf{R} = \mathbf{F} + \mathbf{P} + 80\cos 50°\,\mathbf{i} + 80\sin 50°\,\mathbf{j}\ \text{lb}$$

$$25\,\mathbf{j} = -F\,\mathbf{j} - P\cos 40°\,\mathbf{i} + P\sin 40°\,\mathbf{j} + 80\cos 50°\,\mathbf{i} + 80\sin 50°\,\mathbf{j}$$

$$25\,\mathbf{j} = -F\,\mathbf{j} - 0.766P\,\mathbf{i} + 0.643P\,\mathbf{j} + 51.42\,\mathbf{i} + 61.28\,\mathbf{j}$$

$$\rightarrow 25\,\mathbf{j} = (51.42 - 0.766P)\,\mathbf{i} + (61.28 + 0.643P - F)\,\mathbf{j}$$

Setting the **i** and **j** components on each side of this equation equal gives two scalar equations as

i direction: $0 = 51.42 - 0.766P$

j direction: $25 = 61.28 + 0.643P - F$

Solving these equations simultaneously for P and F gives

$P = 67.1\ \text{lb}$

$F = 79.4\ \text{lb}$

Example 1.2

Calculate the magnitude of **P** and the angles θ and ϕ (as defined in Figure 1.5) for the force $\mathbf{P} = 200\,\mathbf{i} - 500\,\mathbf{j} + 700\,\mathbf{k}\ \text{N}$.

Solution:

The magnitude of the force is

$$P = |\mathbf{P}| = \sqrt{(P_x)^2 + (P_y)^2 + (P_y)^2}$$

$$\rightarrow P = \sqrt{(200)^2 + (-500)^2 + (700)^2} = 883.2 \text{ N}$$

Knowing the magnitude of **P**, the angle θ can be calculated from the expression

$$P_y = P \sin \theta$$

$$-500 = 883.2 \sin \theta$$

$$\rightarrow \theta = \sin^{-1}\left(\frac{-500}{883.2}\right) = -34.5^\circ$$

Similarly, the angle ϕ can be found from the expression

$$P_x = P \cos \theta \cos \phi$$

$$200 = 883.2 \cos(-34.5) \cos \phi$$

$$\rightarrow \phi = \cos^{-1}\left(\frac{200}{(883.2)\cos(-34.5^\circ)}\right) = 74.1^\circ$$

Example 1.3

Find the unit vector in the direction of a line that passes through two points whose coordinates are $(-5,3)$ and $(5,2)$.

Solution:

$$\mathbf{u} = \frac{(x_B - x_A)\,\mathbf{i} + (y_B - y_A)\,\mathbf{j}}{\sqrt{(x_B - x_A)^2 + (y_B - y_A)^2}}$$

$$\rightarrow \mathbf{u} = \frac{(5-(-5))\,\mathbf{i} + (2-3)\,\mathbf{j}}{\sqrt{(5-(-5))^2 + (2-3)^2}} = \frac{10\,\mathbf{i} - \mathbf{j}}{\sqrt{101}} = 0.995\,\mathbf{i} - 0.100\,\mathbf{j}$$

Problems

1.1 Determine the magnitude and direction of the resultant of a 93.2 lb force in the x-direction and a 111.1 lb force in the y-direction.
(Ans. R = 145.0 lb, θ = 50.0° CCW from the positive x-axis)

1.2 Calculate the resultant of the forces acting on the eyebolt below.
(Ans. **R** = 266.6 **i** + 33.2 **j** N)

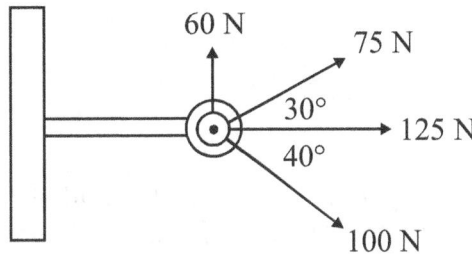

1.3 For a force of **P** = 350 **i** + 700 **j** + 650 **k** N, compute the magnitude of **P** and the angles θ and φ as defined in Figure 1.5.
(Ans. P = 1,017.3 N, θ = 43.5°, φ = 61.7°)

1.4 A force is given as **P** = −125 **i** + 280 **j** − 310 **k** N. Find the magnitude of **P** and the unit vector **u** in the direction of **P**.
(Ans. P = 436.0 N, **u** = −0.287 **i** + 0.642 **j** − 0.711 **k**)

1.5 The magnitude of the force shown below is 175 N. Calculate the unit vector **u** in the direction of the force and write **F** as a vector.
(Ans. **u** = 0.832 **i** + 0.555 **j**, **F** = 145.6 **i** + 97.1 **j** N)

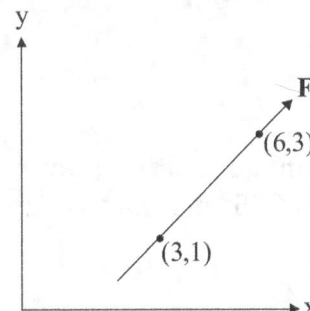

1.6 The resultant of the system of forces shown below is $\mathbf{R} = 425\,\mathbf{j}$ lb. Determine the possible values for the magnitude of $\mathbf{P}$ and the angle θ.
(Ans. $P = 544.8$ and 53.9 lb, $\theta = 33.2°$ and $-87.4°$)

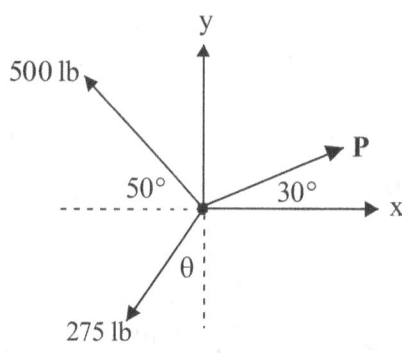

1.7 Three forces applied to an eyebolt are $\mathbf{F}_1 = 100\,\mathbf{i} + 200\,\mathbf{j}$ N, $\mathbf{F}_2 = -500\,\mathbf{i} + 300\,\mathbf{j}$ N, and $\mathbf{F}_3 = 600\,\mathbf{i} + 100\,\mathbf{j}$ N. Compute the resultant force and the angle θ it makes with the horizontal, and write the resultant as the product of a magnitude and a unit vector.
(Ans. $\mathbf{R} = 200\,\mathbf{i} + 600\,\mathbf{j}$ N, $\theta = 71.6°$, $\mathbf{R} = R\,\mathbf{u} = (632.5)(0.316\,\mathbf{i} + 0.949\,\mathbf{j})$ N)

1.8 In the figure below, θ has a value of $14°$. If the resultant of the system of forces is $\mathbf{R} = 500\,\mathbf{j}$ lb, calculate the values of F and P.
(Ans. $F = 701.8$ lb, $P = 579.1$ lb)

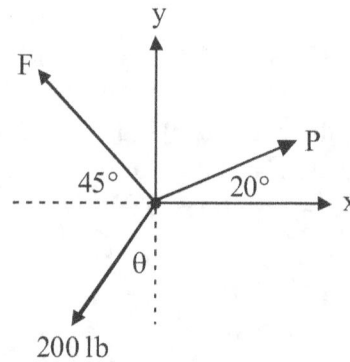

1.9 The resultant of two forces, one in the positive x-direction and the other in the positive y-direction, is 120 lb at an angle of $55°$ measured counterclockwise from the positive x-axis. Determine the two forces.
(Ans. $\mathbf{F}_1 = 68.8\,\mathbf{i}$ lb, $\mathbf{F}_2 = 98.3\,\mathbf{j}$ lb)

1.10 For a force given as $\mathbf{P} = 400\,\mathbf{i} + 600\,\mathbf{j} + 800\,\mathbf{k}$ N, compute the magnitude of $\mathbf{P}$ and determine the angles θ and ϕ as defined in Figure 1.5.
(Ans. $P = 1{,}077.0$ N, $\theta = 33.9°$, $\phi = 63.4°$)

1.11 A force is given as $\mathbf{P} = 200\,\mathbf{i} - 175\,\mathbf{j} + 380\,\mathbf{k}$ N. Find the magnitude of $\mathbf{P}$ and the unit vector $\mathbf{u}$ in the direction of $\mathbf{P}$.
(Ans. P = 463.7 N, $\mathbf{u}$ = 0.431 $\mathbf{i}$ − 0.377 $\mathbf{j}$ + 0.819 $\mathbf{k}$)

1.12 The magnitude of the force shown is 200 N. Determine the unit vector $\mathbf{u}$ in the direction of the force and write $\mathbf{F}$ as a vector.
(Ans. $\mathbf{u}$ = 0.707 $\mathbf{i}$ + 0.707 $\mathbf{j}$, $\mathbf{F}$ = 141.4 $\mathbf{i}$ + 141.4 $\mathbf{j}$ N)

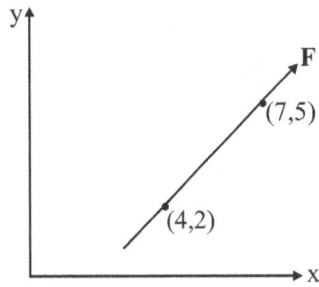

1.13 The resultant of the system of forces shown below is $\mathbf{R} = 387\,\mathbf{j}$ lb. Calculate the possible values of P and θ.
(Ans. P = 522.3 and 164.4 lb, θ = 43.4° and −84.3°)

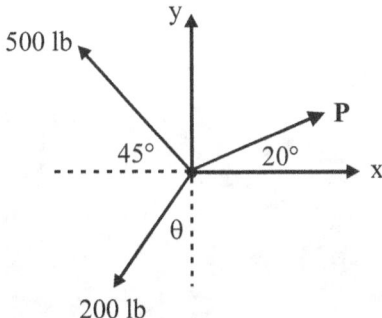

1.14 In Problem 1.13, if the 500 lb force has an unknown magnitude if F, and θ has a value of 14°, find the magnitudes of forces F and P.
(Ans. F = 584.5 lb, P = 491.0 lb)

1.15 A structure is loaded with the three forces: $\mathbf{F}_1 = 200\,\mathbf{i} + 125\,\mathbf{j}$ N, $\mathbf{F}_2 = 300\,\mathbf{i} - 400\,\mathbf{j}$ N, and $\mathbf{F}_3 = -400\mathbf{i} + 100\,\mathbf{j}$ N. Compute the resultant force and the angle θ it makes with the horizontal. Write the resultant as the product of a magnitude and a unit vector.
(Ans. $\mathbf{R}$ = 100.0 $\mathbf{i}$ −175.0 $\mathbf{j}$ N, θ = −60.3°, $\mathbf{R}$ = 201.6(0.496 $\mathbf{i}$ − 0.868 $\mathbf{j}$) N)

(This page was intentionally left blank.)

MODULE 2: Vector Operations

Consider two three-dimensional vectors $\mathbf{P}$ and $\mathbf{Q}$, where θ is the angle between them. The scalar product, or dot product of $\mathbf{P}$ and $\mathbf{Q}$ is defined as

$$\mathbf{P} \cdot \mathbf{Q} = \mathbf{Q} \cdot \mathbf{P} = PQ \cos \theta$$

$$\mathbf{P} \cdot \mathbf{Q} = \left(P_x\, \mathbf{i} + P_y\, \mathbf{j} + P_z\, \mathbf{k}\right) \cdot \left(Q_x\, \mathbf{i} + Q_y\, \mathbf{j} + Q_z\, \mathbf{k}\right) = P_x Q_x + P_y Q_y + P_z Q_z$$

$$\rightarrow PQ \cos \theta = P_x Q_x + P_y Q_y + P_z Q_z$$

where

$$\mathbf{i} \cdot \mathbf{i} = \mathbf{j} \cdot \mathbf{j} = \mathbf{k} \cdot \mathbf{k} = 1$$

$$\mathbf{i} \cdot \mathbf{j} = \mathbf{j} \cdot \mathbf{i} = \mathbf{i} \cdot \mathbf{k} = \mathbf{k} \cdot \mathbf{i} = \mathbf{j} \cdot \mathbf{k} = \mathbf{k} \cdot \mathbf{j} = 0$$

The scalar product can also be used to find the 'projection' of a vector in a certain direction. For example, the vector components of $\mathbf{P}$ can be determined by finding the projections of $\mathbf{P}$ in the $\mathbf{i}, \mathbf{j}$, and $\mathbf{k}$ directions using the relationships

$$P_x = \mathbf{P} \cdot \mathbf{i}$$

$$P_y = \mathbf{P} \cdot \mathbf{j}$$

$$P_z = \mathbf{P} \cdot \mathbf{k}$$

The vector product, or cross product of $\mathbf{P}$ and $\mathbf{Q}$ is defined as

$$\mathbf{P} \times \mathbf{Q} = -\mathbf{Q} \times \mathbf{P} = PQ \sin \theta\, \mathbf{u}$$

$$\begin{aligned} \mathbf{P} \times \mathbf{Q} &= \left(P_x\, \mathbf{i} + P_y\, \mathbf{j} + P_z\, \mathbf{k}\right) \times \left(Q_x\, \mathbf{i} + Q_y\, \mathbf{j} + Q_z\, \mathbf{k}\right) \\ &= \left(P_y Q_z - P_z Q_y\right) \mathbf{i} + \left(P_z Q_x - P_x Q_z\right) \mathbf{j} + \left(P_x Q_y - P_y Q_x\right) \mathbf{k} \end{aligned}$$

$$\rightarrow PQ \sin \theta\, \mathbf{u} = \left(P_y Q_z - P_z Q_y\right) \mathbf{i} + \left(P_z Q_x - P_x Q_z\right) \mathbf{j} + \left(P_x Q_y - P_y Q_x\right) \mathbf{k}$$

where $\mathbf{u}$ is a unit vector perpendicular to both $\mathbf{P}$ and $\mathbf{Q}$, and

$$\mathbf{i} \times \mathbf{i} = \mathbf{j} \times \mathbf{j} = \mathbf{k} \times \mathbf{k} = 0$$

$$\mathbf{i} \times \mathbf{j} = \mathbf{k}$$

$$j \times k = i$$

$$k \times i = j$$

$$j \times i = -k$$

$$k \times j = -i$$

$$i \times k = -j$$

The vector product of any two vectors always yields a vector that is perpendicular (or normal) to both vectors. For a vector product of two-dimensional or coplanar vectors, the result is simply a vector in the z-direction.

Units

Engineering mechanics deals primarily with four fundamental quantities: length, mass, force, and time. The measurement of these quantities can be described in different systems of units. Two common systems used in science and engineering and used in this text are the International System of Units (SI), or metric system, and the U. S. Customary System (US), or English system. Table 1.1 provides the units that are used for each of the four quantities in both the SI and US systems.

Table 2.1 Systems of Units

Quantity	SI	US
mass	kilograms, kg	slugs
length	meters, m	feet, ft
force	newtons, N	pounds, lb
time	seconds, s	seconds, sec

Units in any system must satisfy the requirements of Newton's Second Law (discussed later), which states that force is equal to the mass times the acceleration, given by

$$F = ma$$

Therefore, in SI units, this equivalence shows

$$N = \frac{kg-m}{s^2}$$

and in US units, the equivalence is

$$lb = \frac{slugs-ft}{sec^2}$$

In US units, values of mass may occasionally be provided in units of pounds mass, lb_m, rather than slugs. To avoid conversion difficulties in this case, the best practice is to convert pounds mass to slugs prior to making any calculations, using the conversions

$$1 \text{ slug} = 32.174 \text{ lb}_m$$

$$1 \text{ lb}_m = 0.031 \text{ slug}$$

All subsequent calculations should then be performed using units of slugs for mass rather than pounds mass.

Other useful conversions are

$$1 \text{ ft} = 0.3028 \text{ m}$$

$$1 \text{ slug} = 14.594 \text{ kg}$$

$$1 \text{ lb} = 4.4482 \text{ N}$$

Weight

Since weight is a force, the weight vector can be represented by mass times the gravitational acceleration in the downward direction. In Cartesian coordinates, where the positive y-axis is defined upwards, the proper representation of the weight vector is

$$\mathbf{W} = -mg\,\mathbf{j}$$

Standard values for the acceleration of gravity, g, which are set at sea level and a latitude of 45°, are given as

$$\text{SI units: } g = 9.807 \frac{m}{s^2}$$

$$\text{US units: } g = 32.174 \frac{ft}{sec^2}$$

Approximate values of $9.81 \frac{m}{s^2}$ and $32.2 \frac{ft}{sec^2}$ are generally adequate for most engineering calculations. When weight is discussed, only the magnitude of the weight is usually mentioned since the downward direction is assumed.

Example 2.1

Determine the projection of the force $P = 12\,i - 7\,j$ N, in the direction of the unit vector $u = 0.995\,i - 0.100\,j$.

Solution:

$$P \cdot u = (12\,i - 7\,j) \cdot (0.995\,i - 0.100\,j) = (12)(0.995) + (-7)(-0.100) = 12.6 \text{ N}$$

Example 2.2

Given the vectors $P = 10\,i - 20\,j + 30\,k$ ft, and $Q = -7\,i + 3\,j - 15\,k$ ft, calculate $P \cdot Q$, $P \times Q$, and the angle between P and Q.

Solution:

$$P \cdot Q = (10\,i - 20\,j + 30\,k) \cdot (-7\,i + 3\,j - 15\,k)$$

$$\rightarrow P \cdot Q = (10)(-7) + (-20)(3) + (30)(-15) = -70 - 60 - 450 = -580 \text{ ft}^2$$

$$P \times Q = (10\,i - 20\,j + 30\,k) \times (-7\,i + 3\,j - 15\,k)$$

$$P \times Q = (10)(3)(i \times j) + (10)(-15)(i \times k) + (-20)(-7)(j \times i)$$
$$+(-20)(-15)(j \times k) + (30)(-7)(k \times i) + (30)(3)(k \times j)$$

$$P \times Q = 30\,k - 150(-j) + 140(-k) + 300\,i - 210\,j + 90(-i)$$

$$\rightarrow P \times Q = (300 - 90)\,i + (150-210)\,j + (30-140)\,k = 210\,i - 60\,j - 110\,k \text{ ft}^2$$

Since

$$P \cdot Q = PQ \cos \theta$$

$$P = |P| = \sqrt{(P_x)^2 + (P_x)^2 + (P_x)^2} = \sqrt{(10)^2 + (-20)^2 + (30)^2} = 37.4 \text{ ft}$$

$$Q = |Q| = \sqrt{(Q_x)^2 + (Q_x)^2 + (Q_x)^2} = \sqrt{(-7)^2 + (3)^2 + (-15)^2} = 16.8 \text{ ft}$$

$$\theta = \cos^{-1}\left[\frac{P \cdot Q}{PQ}\right]$$

$$\rightarrow \theta = \cos^{-1}\left[\frac{-580}{(37.42)(16.82)}\right] = 157.2^\circ$$

Example 2.3

Find the projection of $\mathbf{P} = 10\,\mathbf{i} - 8\,\mathbf{j} + 14\,\mathbf{k}$ lb in the direction of a line that passes through the points $(2,-5,1)$ and $(5,2,-4)$.

Solution:

The unit vector passing through the designated points is found by

$$\mathbf{u} = \frac{(x_B - x_A)\,\mathbf{i} + (y_B - y_A)\,\mathbf{j} + (z_B - z_A)\,\mathbf{k}}{\sqrt{(x_B - x_A)^2 + (y_B - y_A)^2 + (z_B - z_A)^2}}$$

$$\rightarrow \mathbf{u} = \frac{(5-2)\,\mathbf{i} + (2-(-5))\,\mathbf{j} + (-4-1)\,\mathbf{k}}{\sqrt{(5-2)^2 + (2-(-5))^2 + (-4-1)^2}} = \frac{3\mathbf{i} + 7\mathbf{j} - 5\mathbf{k}}{\sqrt{83}} = 0.329\,\mathbf{i} + 0.768\,\mathbf{j} - 0.549\,\mathbf{k}$$

The projection of $\mathbf{P}$ in the $\mathbf{u}$ direction is

$$\mathbf{P} \cdot \mathbf{u} = (10\,\mathbf{i} - 8\,\mathbf{j} + 14\,\mathbf{k}) \cdot (0.329\,\mathbf{i} + 0.768\,\mathbf{j} - 0.549\,\mathbf{k})$$

$$\rightarrow \mathbf{P} \cdot \mathbf{u} = (10)(0.329) + (-8)(0.768) + (14)(-0.549) = -10.5 \text{ lb}$$

Problems

2.1 Given the vectors $\mathbf{P} = 8\,\mathbf{i} + 5\,\mathbf{j} - P_z\,\mathbf{k}$ m, and $\mathbf{Q} = 3\,\mathbf{i} - 4\,\mathbf{j} + 2\,\mathbf{k}$ m, determine the value of P_z so that the scalar product of the two vectors is 60 m^2.
(Ans. $P_z = -28$ m)

2.2 Find the projection of the force $\mathbf{P} = 8\,\mathbf{i} - 12\,\mathbf{j} - 28\,\mathbf{k}$ N, in the direction of a line drawn from point $(3,1,-1)$ through point $(-1,2,5)$.
(Ans. $P_u = -29.1$ N)

2.3 Given two vectors $\mathbf{P} = 8\,\mathbf{i} - 14\,\mathbf{j} + 24\,\mathbf{k}$ lb, and $\mathbf{Q} = 40\,\mathbf{i} - 45\,\mathbf{j} + 30\,\mathbf{k}$ lb, calculate the quantities $\mathbf{P} \cdot \mathbf{Q}$, $\mathbf{P} \times \mathbf{Q}$, and the angle between $\mathbf{P}$ and $\mathbf{Q}$.
(Ans. $\mathbf{P} \cdot \mathbf{Q} = 1{,}670$ lb^2, $\mathbf{P} \times \mathbf{Q} = 660\,\mathbf{i} + 720\,\mathbf{j} + 200\,\mathbf{k}$ lb^2, $\theta = 30.8°$)

2.4 Given the vectors $\mathbf{P} = 7\,\mathbf{i} + P_y\,\mathbf{j}$ m and $\mathbf{Q} = 9\,\mathbf{i} + Q_y\,\mathbf{j}$ m. Find the possible values of P_y and Q_y if $\mathbf{P} \times \mathbf{Q}$ is $20\,\mathbf{k}$ m^2 and $\mathbf{P} \cdot \mathbf{Q}$ is 75 m^2.
(Ans. $P_y = 2.1$ m, -4.4 m, $Q_y = 5.6$ m, -2.8 m)

2.5 Two vectors are given as $\mathbf{P} = 20\,\mathbf{i} + P_y\,\mathbf{j} + 30\,\mathbf{k}$ ft and $\mathbf{Q} = Q_x\,\mathbf{i} + 40\,\mathbf{j} + Q_z\,\mathbf{k}$ ft. If $\mathbf{P} \times \mathbf{Q} = -1{,}100\,\mathbf{i} + 1{,}000\,\mathbf{j} - 600\,\mathbf{k}$ ft^2, determine the values of P_y, Q_x, and Q_z.
(Ans. $P_y = 40.0$ ft, $Q_x = 35.0$ ft, $Q_z = 2.5$ ft)

2.6 Three vectors are given as $P = 4\,i - 6\,j + 8\,k$ N, $Q = 8\,i - j + 5\,k$ N, and $R = -2\,i + 5\,j + 4\,k$ N. Calculate $(P \times Q) \times R$, $P \times (Q \times R)$, $(P \times Q) \cdot R$, and $(Q \times R) \cdot P$. (Ans. $(P \times Q) \times R = -44\,i - 22\,k$ N^3, $P \times (Q \times R) = 108\,i - 384\,j - 342\,k$ N^3, $(P \times Q) \cdot R = 440$ N^3, $(Q \times R) \cdot P = 440$ N^3)

2.7 Three force vectors are given as $P = 4\,i + 8\,j - k$ N, $Q = -6\,i - 7\,j + 2\,k$ N, and $R = 10\,i + 5\,j + 9\,k$ N. Compute the vector triple products $Q \cdot (P \times R)$, $P \cdot (R \times Q)$, and $R \cdot (Q \times P)$. (Ans. $Q \cdot (P \times R) = -260$ N^3, $P \cdot (R \times Q) = -260$ N^3, $R \cdot (Q \times P) = -260$ N^3)

2.8 Two position vectors are given as $P = P_x\,i + 45\,j + P_z\,k$ m and $Q = 40\,i + Q_y\,j + 15\,k$ m. If $P \times Q = 700\,i - 75\,j - 1{,}200\,k$ m^2, determine the values of P_x, P_z, and Q_y. (Ans. $P_x = 4.5$ m, $P_z = -0.2$ m, $Q_y = 133.3$ m)

2.9 Given the vectors $P = 7\,i - 4\,j + P_z\,k$ m, and $Q = 4\,i + 2\,j - 1\,k$ m, find the value of P_z so that the scalar product of the two vectors will be 50 m^2. (Ans. $P_z = -30$ m)

2.10 Determine the projection of the force $P = 10\,i - 20\,j - 30\,k$ N, in the direction of a line drawn from point $(2,1,-1)$ through point $(0,1,6)$. (Ans. $P_u = -31.6$ N)

2.11 Given two vectors $P = 10\,i + 20\,j + 30\,k$ lb, and $Q = 50\,i - 40\,j - 25\,k$ lb, calculate the quantities $P \cdot Q$, $P \times Q$, and the angle between P and Q. (Ans. $P \cdot Q = -1{,}050$ lb^2, $P \times Q = 700\,i + 1{,}750\,j - 1{,}400\,k$ lb^2, $\theta = 114.1°$)

2.12 Given the vectors $P = 8\,i + P_y\,j$ m and $Q = 15\,i - Q_y\,j$ m. Compute the possible values of P_y and Q_y if $P \times Q = -10\,k$ m^2 and $P \cdot Q = 135$ m^2. (Ans. $P_y = -2.5$, 3.2 m, $Q_y = 6.0$, -4.7 m)

2.13 Two vectors are given as $P = 30\,i + P_y\,j + 20\,k$ ft and $Q = Q_x\,i + 50\,j + Q_z\,k$ ft. If $P \times Q = -1{,}125\,i + 1{,}050\,j - 500\,k$ ft^2, find the values of P_y, Q_x, and Q_z. (Ans. $P_y = 41.7$ ft, $Q_x = 48.0$ ft, $Q_z = -3.0$ ft)

2.14 Three vectors are given as $P = 5\,i - 7\,j - 9\,k$ N, $Q = 7\,i + 2\,j + 6\,k$ N, and $R = -3\,i + 4\,j + 3\,k$ N. Determine $(P \times Q) \times R$, $P \times (Q \times R)$, $(P \times Q) \cdot R$, and $(Q \times R) \cdot P$. (Ans. $(P \times Q) \times R = -515\,i - 105\,j - 375\,k$ N^3, $P \times (Q \times R) = -589\,i - 8\,j - 321\,k$ N^3, $(P \times Q) \cdot R = -123$ N^3, $(Q \times R) \cdot P = -123$ N^3)

2.15 Given that $P = 6\,i + P_y\,j$ m, $Q = 20\,i - Q_y\,j$ m, $P \times Q = -25\,k$ m^2, and $P \cdot Q = 200$ m^2 Calculate all the possible values for P_y and Q_y. (Ans. $P_y = -4.3$, 5.5 m, $Q_y = 18.4$, -14.3 m)

MODULE 3: Moment of a Force

While a force has the tendency to move a body in the direction in which the force is applied, it may also rotate the body about an axis that does not intersect the line-of-action of the force. This tendency to rotate the body about an axis is known as the moment **M** of the force. In Figure 3.1, the two-dimensional force **F** is applied such that its line-of-action passes through point A. Since the line-of-action of **F** does not pass through the origin O of the coordinate system, **F** creates a rotation about an axis that passes through point O and is perpendicular to both the x- and y-axes.

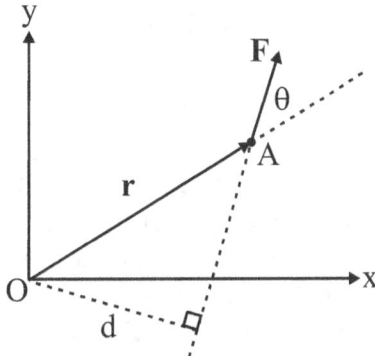

Figure 3.1

The moment **M** created about the rotational axis is defined by the vector product

$$\mathbf{M} = \mathbf{r} \times \mathbf{F}$$

where **r** is the position vector from O to <u>any</u> point on the line-of-action of **F**. It's important to note that the order of **r** × **F** must be maintained because **F** × **r** would produce a moment having the opposite sense, resulting in **F** × **r** = −**M** instead. Since the moment is equal to the vector product of **r** and **F**, the magnitude of the moment M can be expressed as the scalar quantity

$$M = |\mathbf{M}| = Fr \sin \theta$$

Often the perpendicular distance d between O and the line-of-action of **F** is used in this formula, which corresponds to the case where $\theta = 90°$, therefore $\sin \theta = 1.0$ and

$$M = Fd$$

In the general three-dimensional case, the moment of **F** about an axis through point O as

$$\mathbf{M} = \mathbf{r} \times \mathbf{F} = (x\,\mathbf{i} + y\,\mathbf{j} + z\,\mathbf{k}) \times (F_x\,\mathbf{i} + F_y\,\mathbf{j} + F_z\,\mathbf{k})$$
$$= (yF_z - zF_y)\,\mathbf{i} + (zF_x - xF_z)\,\mathbf{j} + (xF_y - yF_x)\mathbf{k}$$

$$\rightarrow \mathbf{M} = M_x\,\mathbf{i} + M_y\,\mathbf{j} + M_z\,\mathbf{k}$$

where

$$M_x = yF_z - zF_y$$

$$M_y = zF_x - xF_z$$

$$M_z = xF_y - yF_x$$

Here, M_x, M_y, and M_z are called the components of the moment **M**. The units of a moment are force times distance, so in the SI system a moment has units of N-m and in the US system a moment has units of lb-ft. The sense or direction of a moment can be determined by the 'right hand' rule. This rule states that when the fingers of the right hand are curled in the direction of the tendency to rotate, the thumb will point in the direction of the moment. For example, in a coplanar case when the rotation is counterclockwise, the direction of the moment will be perpendicular to the xy-plane and pointing in the positive z-direction. This sense is used to designate a positive moment. If the tendency to rotate is clockwise, the direction of the moment will be in the negative z-direction. This is used to designate a negative moment. If a single force acts at a location on a body that creates a tendency to rotate the body about a particular axis, an equivalent system of loads comprising a force applied at the rotation axis and a moment applied about the rotation axis can be determined. Here the applied moment is calculated by

$$\mathbf{M} = \mathbf{r} \times \mathbf{F}$$

These equivalent systems are shown in Figure 3.2.

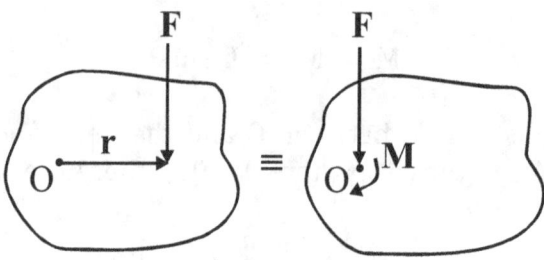

Figure 3.2

The projection of a moment in particular directions, i.e., about specified axes, can also be determined using a scalar product as was done for forces. For example, the projection of moment **M** in the same direction as unit vector **u**, can be determined by

$$M_u = \mathbf{M} \cdot \mathbf{u}$$

For a three-dimensional moment, the vector components of this moment in the x-, y-, and z-directions can be found using the relationships

$$M_x = \mathbf{M} \cdot \mathbf{i}$$

$$M_y = \mathbf{M} \cdot \mathbf{j}$$

$$M_z = \mathbf{M} \cdot \mathbf{k}$$

Example 3.1

For **r** = 5 **i** + 3 **j** m, and **F** = 100 **i** + 150 **j** N, compute the moment created by **F** about the origin of the coordinate system.

Solution:

$$\mathbf{M} = \mathbf{r} \times \mathbf{F}$$

$$\rightarrow \mathbf{M} = (5\,\mathbf{i} + 3\,\mathbf{j}) \times (100\,\mathbf{i} + 150\,\mathbf{j}) = (5)(150)\,\mathbf{k} + (3)(100)(-\mathbf{k}) = 450\,\mathbf{k}\text{ N-m}$$

Example 3.2

A force is given as **F** = 68.7 **i** + 22.9 **j** – 50.2 **k** N and is acting through point B (5,1,1) m. Determine the moment of this force about point C (3,4,−3) m.

Solution:

The moment about point C is calculated using a position vector from C to any point on the line-of-action of **F**. The position vector from C to B can therefore be written as

$$\mathbf{r} = (x_B - x_C)\,\mathbf{i} + (y_B - y_C)\,\mathbf{j} + (z_B - z_C)\,\mathbf{k}$$

$$\rightarrow \mathbf{r} = (5 - 3)\,\mathbf{i} + (1 - 4)\,\mathbf{j} + (1 - (-3))\,\mathbf{k} = 2\,\mathbf{i} - 3\,\mathbf{j} + 4\,\mathbf{k}\text{ m}$$

The moment of **F** about point C is then

$$\mathbf{M_C} = \mathbf{r} \times \mathbf{F} = (2\,\mathbf{i} - 3\,\mathbf{j} + 4\,\mathbf{k}) \times (68.7\,\mathbf{i} + 22.9\,\mathbf{j} - 50.2\,\mathbf{k})$$

$$\mathbf{M_C} = (2)(22.9)\,\mathbf{k} + (2)(-50.2)(-\mathbf{j}) + (-3)(68.7)(-\mathbf{k}) + (-3)(-50.2)\,\mathbf{i} \\ + (4)(68.7)\,\mathbf{j} + (4)(22.9)(-\mathbf{i})$$

$$\rightarrow \mathbf{M_C} = [(-3)(-50.2) - (4)(22.9)]\,\mathbf{i} + [-(2)(-50.2) + (4)(68.7)]\,\mathbf{j} \\ + [(2)(22.9) - (-3)(68.7)]\,\mathbf{k} = 59.0\,\mathbf{i} + 375.2\,\mathbf{j} + 251.9\,\mathbf{k} \text{ N-m}$$

Example 3.3

Calculate the projection of the moment of the force $\mathbf{F} = 3\,\mathbf{i} + 4\,\mathbf{j} - 5\,\mathbf{k}$ N, acting through point A, having coordinates (4,1,1), in the same direction as a line passing from point B (2,5,−2) to point C (4,−1,1).

Solution:

First, the moment about a point on line BC must be determined. This can be done with respect to either point B or point C. Using point B, the position vector from B to A is

$$\mathbf{r} = (x_A - x_B)\,\mathbf{i} + (y_A - y_B)\,\mathbf{j} + (z_A - z_B)\,\mathbf{k}$$

$$\rightarrow \mathbf{r} = (4 - 2)\,\mathbf{i} + (1 - 5)\,\mathbf{j} + (1 - (-2))\,\mathbf{k} = 2\,\mathbf{i} - 4\,\mathbf{j} + 3\,\mathbf{k} \text{ m}$$

The moment of **F** about point B is then

$$\mathbf{M} = \mathbf{r} \times \mathbf{F} = (2\,\mathbf{i} - 4\,\mathbf{j} + 3\,\mathbf{k}) \times (3\,\mathbf{i} + 4\,\mathbf{j} - 5\,\mathbf{k})$$

$$\mathbf{M} = 2(4)\,\mathbf{k} + (2)(-5)(-\mathbf{j}) + (-4)(3)(-\mathbf{k}) + (-4)(-5)\,\mathbf{i} + (3)(3)\,\mathbf{j} + (3)(4)(-\mathbf{i})$$

$$\rightarrow \mathbf{M} = 8\,\mathbf{i} + 19\,\mathbf{j} + 20\,\mathbf{k} \text{ N-m}$$

Next, the unit vector in the direction from point B to point C must be found from

$$\mathbf{u} = \frac{(x_B - x_A)\,\mathbf{i} + (y_B - y_A)\,\mathbf{j} + (z_B - z_A)\,\mathbf{k}}{\sqrt{(x_B - x_A)^2 + (y_B - y_A)^2 + (z_B - z_A)^2}}$$

$$\mathbf{u} = \frac{(4 - 2)\,\mathbf{i} + (-1 - 5)\,\mathbf{j} + (1 - (-2))\,\mathbf{k}}{\sqrt{(4 - 2)^2 + (-1 - 5)^2 + (1 - (-2))^2}} = \frac{2\,\mathbf{i} - 6\,\mathbf{j} + 3\,\mathbf{k}}{\sqrt{49}}$$

$$\rightarrow \mathbf{u} = 0.286\,\mathbf{i} - 0.857\,\mathbf{j} + 0.429\,\mathbf{k}$$

Taking the projection of the moment in the direction of **u** gives

$M_u = \mathbf{M} \cdot \mathbf{u}$

$M_u = (8\,\mathbf{i} + 19\,\mathbf{j} + 20\,\mathbf{k}) \cdot (0.286\,\mathbf{i} - 0.857\,\mathbf{j} + 0.429\,\mathbf{k})$

$\rightarrow M_u = (8)(0.286) + (19)(-0.857) + (20)(0.429) = -5.4 \text{ N-m}$

Problems

3.1 Determine the moment of the force $\mathbf{F} = -5\,\mathbf{i} - 2\,\mathbf{j} + 3\,\mathbf{k}$ lb, acting through point A (2,1,−1) ft in the same direction as a line passing from point B (2,−5,−5) ft to point C (0,−3,2) ft.
(Ans. M_u = 15.6 lb-ft)

3.2 A force $\mathbf{F} = 6{,}000\,\mathbf{i} - F_y\,\mathbf{j} + 5{,}000\,\mathbf{k}$ N, is acting through point A (0,0,0) m. If $\mathbf{F}$ creates a force component of −20,000 N in the same direction as a line passing from point B (−2,5,4) m to point C (4,7,3) m, calculate the value of F_y.
(Ans. F_y = 79,621.8 N)

3.3 In the figure below, if M = 200 kg, find the value of P if the sum of the moments about point O is zero.
(Ans. P = 613.1 N)

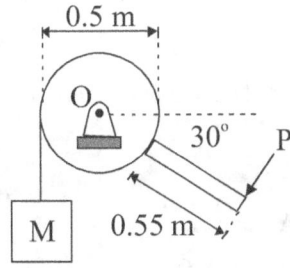

3.4 In the figure above, if P = 766.4 N and the sum of the moments about point O is zero, compute the mass M.
(Ans. M = 250 kg)

3.5 The force $\mathbf{F} = 5\,\mathbf{i} - 6\,\mathbf{j} - 2\,\mathbf{k}$ N is acting through point A (−2,5,−3) m. Determine the component of the moment created by this force along a line from point B (1,3,5) m to point C (2,6,10) m.
(Ans. M_{BC} = −25.4 N-m)

3.6 The plate below comprises 1-ft squares and the forces are applied shown. Calculate the moment about point B created by these forces.
(Ans. $\mathbf{M_B} = -189.3\,\mathbf{k}$ lb-ft.)

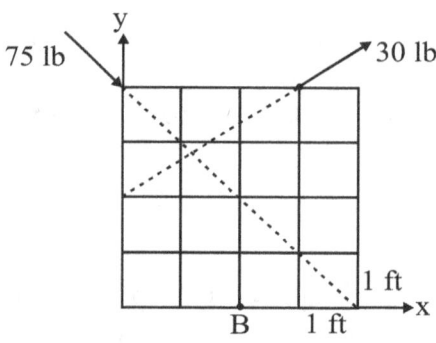

3.7 For the boom shown below, determine the tension in cable AB if the sum of the moments about point O of the boom is zero and $\theta = 41°$.
(Ans. T = 9,609.1 lb)

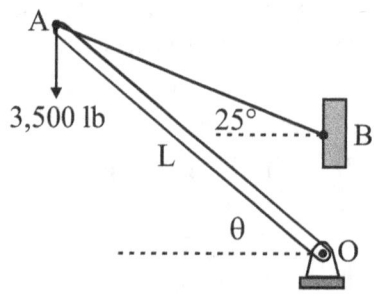

3.8 Find the moment about points O created by the applied force.
(Ans. $\mathbf{M_O} = -5,525.5\,\mathbf{k}$ lb-in)

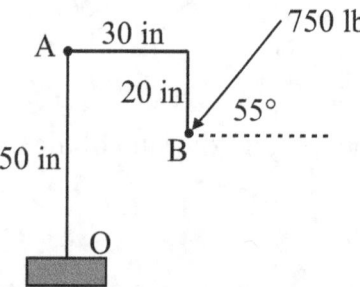

3.9 Calculate the value of F_z if $\mathbf{F} = -300\,\mathbf{i} + 500\,\mathbf{j} + F_z\,\mathbf{k}$ N acts through point A (0,0,0) m and creates a moment component of $-3,000$ N-m in the same direction as a line passing from point B (2,4,5) m to point C (-3,6,8) m.
(Ans. $F_z = -99.5$ N)

3.10 Compute the moment of the force $\mathbf{F} = 2\,\mathbf{i} + 3\,\mathbf{j} + 4\,\mathbf{k}$ lb, acting through point A (3,0,1) ft in the direction of a line passing from point B (−2,5,5) ft to point C (−3,0,1) ft.
(Ans. M_u = 7.4 lb-ft)

3.11 A force $\mathbf{F} = F_x\,\mathbf{i} + 500\,\mathbf{j} + 700\,\mathbf{k}$ N, is acting through point A (0,0,0) m. If $\mathbf{F}$ creates a moment component of −2,000 N-m in the direction of a line passing from point B (2,4,5) m to point C (−3,6,8) m, determine the value of F_x.
(Ans. F_x = −6,737.5 N)

3.12 As shown below, forces are applied to the plate, which is made up of 1-ft squares. Calculate the moment about the origin of the given coordinate system due to these forces.
(Ans. $\mathbf{M_B}$ = 30.0 $\mathbf{k}$ lb-ft.)

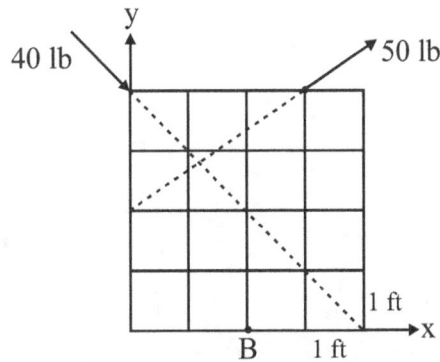

3.13 For the boom shown below, determine the angle θ if the tension in cable AB is 5,083.9 lb and the sum of the moments about point O is zero.
(Ans. θ = 38°)

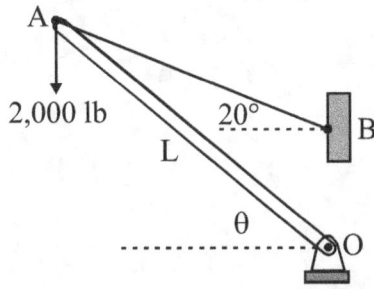

3.14 Calculate the moments about both points O and A due to the applied force.
(Ans. $\mathbf{M}_O = 449.2\ \mathbf{k}$ lb-in, $\mathbf{M}_A = -14{,}977.6\ \mathbf{k}$ lb-in)

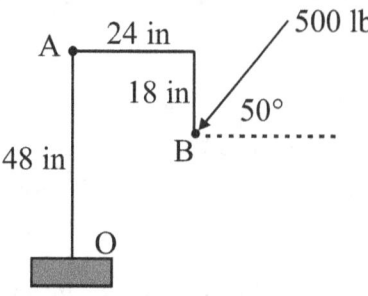

3.15 In Problem 3.11, if $\mathbf{F} = 200\ \mathbf{i} - 600\ \mathbf{j} - F_z\ \mathbf{k}$ N, determine the value of F_z.
(Ans. $F_z = 1{,}307.4$ N)

MODULE 4: Force-Couple Systems

A force acting on a body moves it in the direction the force is applied and may rotate the body about an axis within the body. These effects can better be visualized using couples. A couple is created when two equal forces act in the opposite direction, a distance d apart, as shown in Figure 4.1.

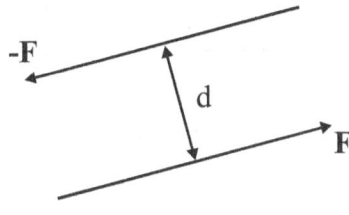

Figure 4.1

Since d is perpendicular to both forces, the magnitude of the couple M is found by

$$M = Fd$$

The direction of the couple is again determined by the right hand rule. A couple applied at any point on a body exerts the same resultant at every point on the body.

Figure 4.2 shows a force **F** acting at point A on the body, which creates a rotation about point B on the body. Forces equal and opposite to **F** can then be applied at point B to form an equivalent system of forces acting on the body. The original force **F** applied at A and −**F** applied at B can then be replaced by a counterclockwise couple **M** as shown.

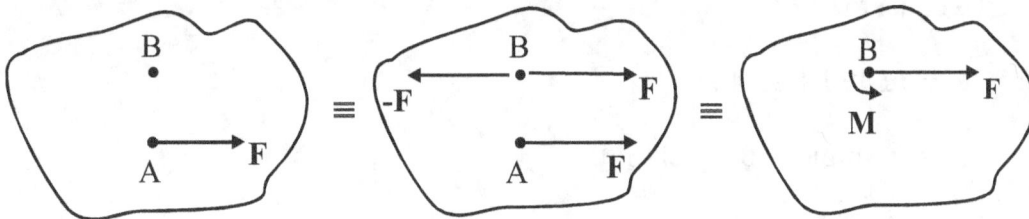

Figure 4.2

Therefore, the original applied force at A has been replaced by an equivalent system comprising a force and a couple at B, without altering the effects of the original force on the body. This is known as a force-couple system. Forming equivalent systems is widely used in engineering mechanics and will be very important in the modules that follow.

Example 4.1

Replace the 15 N force at point A with a force-couple system (a) at point O, and (b) at point B.

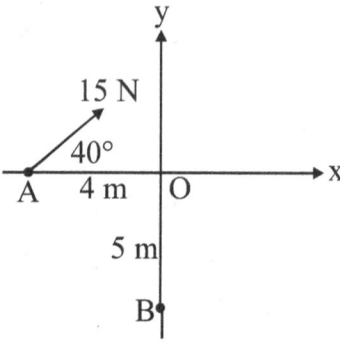

Solution:

(a) The 15 N force applied at A can be replaced with a force and couple applied at point O equal to

$$\mathbf{R} = \mathbf{F} = F \cos \theta \, \mathbf{i} + F \sin \theta \, \mathbf{j}$$

$$\rightarrow \mathbf{R} = \mathbf{F} = 15 \cos 40^\circ \, \mathbf{i} + 15 \sin 40^\circ \, \mathbf{j} = 11.49 \, \mathbf{i} + 9.64 \, \mathbf{j} \text{ N}$$

$$\mathbf{M_O} = \mathbf{r_{OA}} \times \mathbf{F}$$

$$\rightarrow \mathbf{M_O} = -4 \, \mathbf{i} \times (11.49 \, \mathbf{i} + 9.64 \, \mathbf{j}) = -38.6 \, \mathbf{k} \text{ N-m}$$

(b) The 15 N force applied at A can also be replaced with a force applied at point B equal to

$$\mathbf{R} = 11.49 \, \mathbf{i} + 9.64 \, \mathbf{j} \text{ N}$$

plus, a couple at B equal to

$$\mathbf{M_B} = \mathbf{r_{BA}} \times \mathbf{R}$$

$$\mathbf{M_B} = (-4 \, \mathbf{i} + 5 \, \mathbf{j}) \times (11.49 \, \mathbf{i} + 9.64 \, \mathbf{j}) = (-4)(9.64) \, \mathbf{k} + (5)(11.49)(-\mathbf{k})$$

$$\rightarrow \mathbf{M_B} = -96.0 \, \mathbf{k} \text{ N-m}$$

Example 4.2

For the square block loaded as shown, replace the applied force and couple with an equivalent force and a couple at point O.

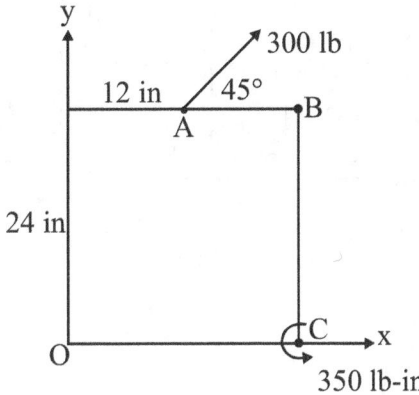

Solution:

The single force placed at O is the resultant of the applied forces calculated by

$$\mathbf{R} = 300 \cos 45° \,\mathbf{i} + 300 \sin 45° \,\mathbf{j} \text{ lb}$$

$$\rightarrow \mathbf{R} = 212.1 \,\mathbf{i} + 212.1 \,\mathbf{j} \text{ lb}$$

The couple placed at point O is equal to the sum of the moments of the applied force and the applied moment calculated by

$$\mathbf{M_O} = \mathbf{r}_{OA} \times (300 \cos 45° \,\mathbf{i} + 300 \sin 45° \,\mathbf{j}) + 350 \,\mathbf{k}$$

$$\mathbf{M_O} = (12 \,\mathbf{i} + 24 \,\mathbf{j}) \times (212.1 \,\mathbf{i} + 212.1 \,\mathbf{j}) + 350 \,\mathbf{k}$$

$$\rightarrow \mathbf{M_O} = (12)(212.1) \,\mathbf{k} + (24)(212.1)\,(-\mathbf{k}) + 350 \,\mathbf{k} = -2{,}195.2 \,\mathbf{k} \text{ lb-in}$$

Example 4.3

Determine the magnitude of the force F so that the moment at the wall is zero

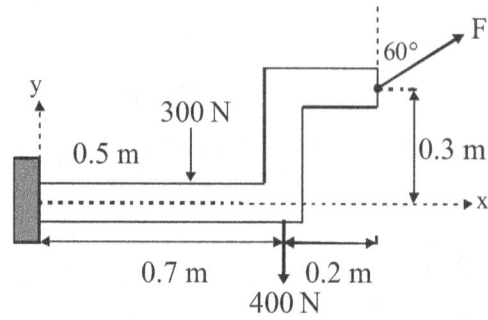

Solution:

$$M = r_1 \times F_1 + r_2 \times F_2 + r \times F$$

$$M = 0.5\,i \times (-300\,j) + 0.7\,i \times (-400\,j) + (0.9\,i + 0.3\,j) \times (F \sin 60°\,i + F \cos 60°\,j)$$

$$M = -150\,k - 280\,k + [(0.9)(0.500) - (0.3)(0.866)]F\,k$$

$$M = -430 + 0.19F\,k$$

$$0 = -430 + 0.19F$$

$$\rightarrow F = 2,263.2\text{ N}$$

Problems

4.1 For the case of F = 250 N, compute the combined moment of these forces about point O.
(Ans. M_O = 6,375.0 k N-m)

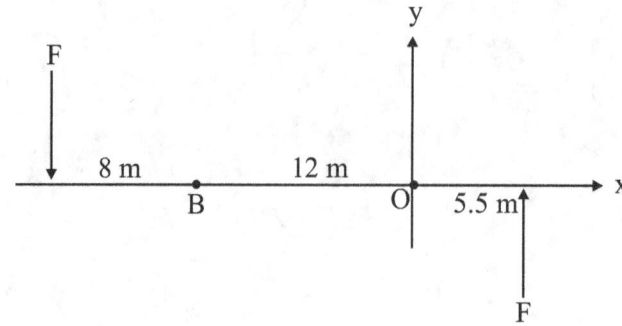

4.2 In the figure above, if the moments created by F about points O and B are both equal to 1,300 N-m, find F.
(Ans. F = 100 N)

4.3 In the figure below, replace the applied forces and moment with an equivalent system of a single force, **R**, and couple **M**$_O$ at the center of the wheel. The applied moment M has a value of 2,500 lb-in and the outside diameter of the wheel is 4.25 ft.
(Ans. **R** = −30.2 **i** + 614.4 **j** lb, **M**$_O$ = 13,166.3 **k** lb-in)

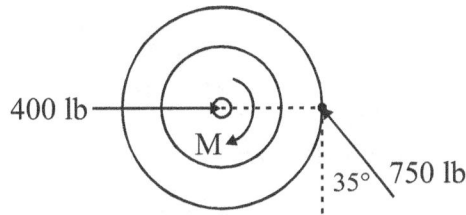

4.4 In Example 4.3, determine the moment at the wall if **F** = 450 **i** + 625 **j** − 300 **k** N.
(Ans. **M**$_A$ = −90.0 **i** +270.0 **j** − 2.5 **k** N-m)

4.5 In the structure shown below, replace the applied forces with a single force and couple applied at point O to form an equivalent system.
(Ans. **R** = 2,500.0 **i** − 3,000.0 **j** N, **M**$_O$ = −1,500.0 **k** N-m)

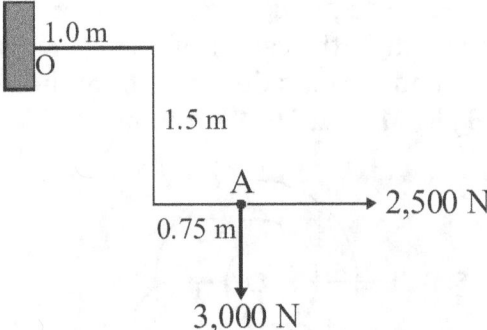

4.6 If F = 550 N, calculate the value of d in the figure below so that the force and couple shown can be replaced by a single force applied at point A to form an equivalent system.
(Ans. d = 0.45 m)

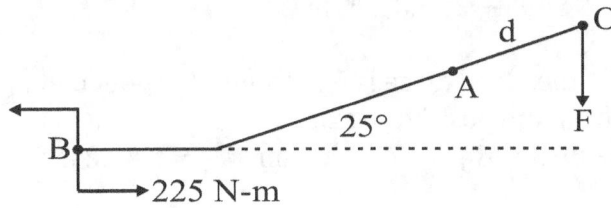

4.7 Solve Example 4.3 for the case where **F** makes an angle of 20° with the horizontal.
(Ans. F = 16,538.5 N)

4.8 In the figure below, determine the magnitude of the force **F** for the case of d = 0.6 m, so
that the force and couple shown can be replaced by a single force applied at point A to
form an equivalent system.
(Ans. F = 531.9 N)

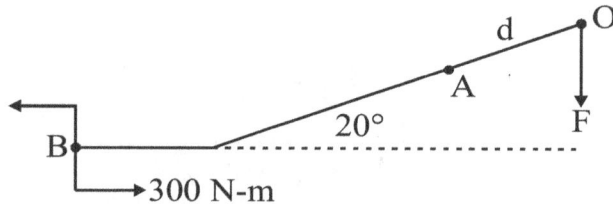

4.9 For Example 4.2, replace the system of a force and a moment with a single force and a
couple at point C.
(Ans. **R** = 212.1 **i** + 212.1 **j** lb, **M**$_C$ = −7,286.3 **k** lb-in)

4.10 From the results of Problem 4.9, find the replacement couple at point B rather than
point C.
(Ans. **M**$_B$ = −2,195.6 **k** lb-in)

4.11 In the wheel below, replace the applied forces and moment with an equivalent system
of a single force, **R**, and couple at the center of the wheel. The applied moment M
has a value of 2,000 lb-in and the outside diameter of the wheel is 7.0 ft.
(Ans. **R** = 100 **i** + 692.8 **j** lb, **M**$_O$ = 27,098.4 **k** lb-in)

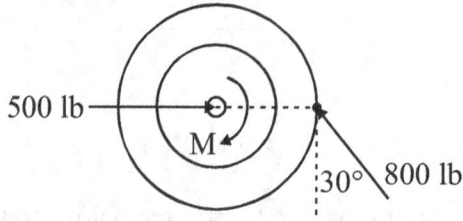

4.12 In Example 4.3, calculate the moment about the wall if F is replaced by the force
F = 866.0 **i** + 500 **j** − 600 N.
(Ans. **M**$_A$ = −180 **i** + 540 **j** − 239.8 **k** N-m)

4.13 Solve Example 4.1 if the given force is 80 N, and it's applied at an angle which is 120°
counterclockwise from the positive x-axis.
(Ans. **R** = −40.0 **I** + 69.3 **j**, **M**$_O$ = 277.1 **k** N-m, **M**$_B$ = −77.2 **k** N-m)

4.14 Determine the value of d in the figure below so that the force and couple shown can be replaced by a single force applied at point A to form an equivalent system. Let F = 400 N.

(Ans. d = 0.4 m)

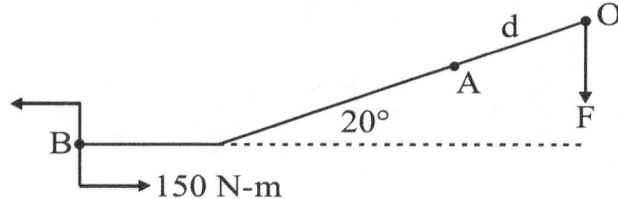

4.15 Solve Example 4.2 for the case where the applied force makes an angle of 30° with the horizontal.

(Ans. $\mathbf{M_O} = -4{,}085.2\ \mathbf{k}$ lb-in)

(This page was intentionally left blank.)

MODULE 5: Coplanar Force Systems

A coplanar system of forces is one in which all forces lie in the same plane. These forces can be concurrent or parallel. A concurrent system of forces is one where the lines-of-action of all the forces intersect at a point, as shown in Figure 5.1.

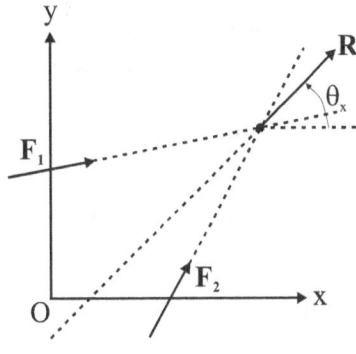

Figure 5.1

A parallel system of forces is one where the lines-of-action of the forces do not intersect, as shown in Figure 5.2.

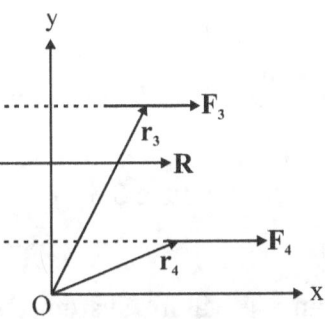

Figure 5.2

The resultant of a combined concurrent and parallel force system is both a resultant force and moment applied at point O. The resultant force is determined by the summation of all the applied forces as

$$R = \sum F = \left(\sum F_x\right)i + \left(\sum F_y\right)j$$

The magnitude and angle of orientation of the resultant force are given by

$$R = |\mathbf{R}| = \sqrt{\left(\sum F_x\right)^2 + \left(\sum F_y\right)^2}$$

where

$$\theta_x = \tan^{-1}\left[\frac{(\sum F_y)}{(\sum F_x)}\right]$$

The resultant moment is determined by the equation

$$\mathbf{M}_O = \sum(\mathbf{r} \times \mathbf{F})$$

A system of forces that includes both concurrent and parallel forces is shown in Figure 5.3.

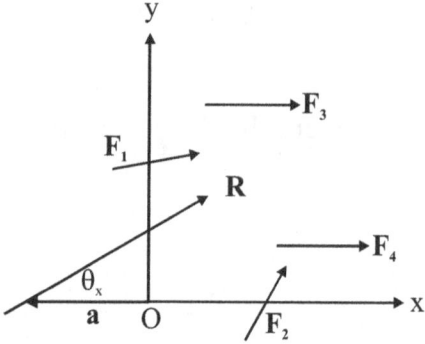

Figure 5.3

This system can be replaced by an equivalent system comprising a single resultant force, located by position vector **a**, so that the resultant force creates a moment about point O equal to the resultant moment. Vector **a**, locating the line-of-action of the resultant force, can be obtained by satisfying the relationship

$$\mathbf{a} \times \mathbf{R} = \mathbf{M}_O = \sum(\mathbf{r} \times \mathbf{F})$$

Any moments that are applied directly to the body, indicated by $\sum \mathbf{M}$, must be added to the moment calculation as

$$\mathbf{a} \times \mathbf{R} = \mathbf{M}_O = \sum(\mathbf{r} \times \mathbf{F}) + \sum \mathbf{M}$$

Example 5.1

For the block loaded as shown below, replace the applied forces and moments with a single force and a couple at point O.

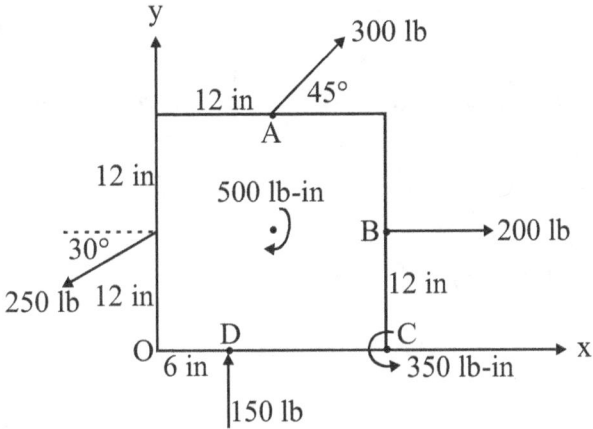

Solution:

The single force to be placed at O will be the resultant of the applied forces calculated by

$$\mathbf{R} = 300 \cos 45° \, \mathbf{i} + 300 \sin 45° \, \mathbf{j} + 200 \, \mathbf{i} + 150 \, \mathbf{j} - 250 \cos 30° \, \mathbf{i} - 250 \sin 30° \, \mathbf{j}$$

$$\rightarrow \mathbf{R} = 212.1 \, \mathbf{i} + 212.1 \, \mathbf{j} + 200 \, \mathbf{i} + 150 \, \mathbf{j} - 216.5 \, \mathbf{i} - 125 \, \mathbf{j} = 195.6 \, \mathbf{i} + 237.1 \, \mathbf{j} \, \text{lb}$$

The couple to be placed at point O will be equal to the sum of the moments of each of the applied forces and the applied moments calculated by

$$\mathbf{M_O} = \mathbf{r_{OA}} \times (300 \cos 45° \, \mathbf{i} + 300 \sin 45° \, \mathbf{j}) + \mathbf{r_{OB}} \times 200 \, \mathbf{i} + \mathbf{r_{OD}} \times 150 \, \mathbf{j}$$
$$+ \mathbf{r_{OE}} \times (-250 \cos 30° \, \mathbf{i} - 250 \sin 30° \, \mathbf{j}) - 500 \, \mathbf{k} + 350 \, \mathbf{k}$$

$$\mathbf{M_O} = (12 \, \mathbf{i} + 24 \, \mathbf{j}) \times (212.1 \, \mathbf{i} + 212.1 \, \mathbf{j}) + (24 \, \mathbf{i} + 12 \, \mathbf{j}) \times 200 \, \mathbf{i} + 6 \, \mathbf{i} \times 150 \, \mathbf{j}$$
$$+ 12 \, \mathbf{j} \times (-216.5 \, \mathbf{i} - 125 \, \mathbf{j}) - 500 \, \mathbf{k} + 350 \, \mathbf{k}$$

$$\mathbf{M_O} = (12)(212.1) \, \mathbf{k} + (24)(212.1)(-\mathbf{k}) + (12)(200)(-\mathbf{k}) + (6)(150) \, \mathbf{k}$$
$$+ (12)(-216.5)(-\mathbf{k}) - 500 \, \mathbf{k} + 350 \, \mathbf{k}$$

$$\rightarrow \mathbf{M_O} = -1{,}597.2 \, \mathbf{k} \, \text{lb-in}$$

The negative sign indicates a <u>clockwise</u> moment about point O.

Example 5.2

For the parallel system of forces shown below, find the resultant force **R** and determine its location by calculating the magnitude of vector **a** relative to point O as shown.

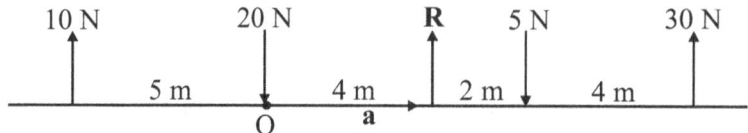

Solution:

$$\mathbf{R} = \sum \mathbf{F} = 10\,\mathbf{j} - 20\,\mathbf{j} - 5\,\mathbf{j} + 30\,\mathbf{j}\ \text{N}$$

$$\rightarrow \mathbf{R} = 15\,\mathbf{j}\ \text{N}$$

$$\mathbf{M_O} = \mathbf{a} \times \mathbf{R} = \sum(\mathbf{r} \times \mathbf{F})$$

$$\rightarrow a\,\mathbf{i} \times 15\,\mathbf{j} = -5\,\mathbf{i} \times 10\,\mathbf{j} + 6\,\mathbf{i} \times (-5\,\mathbf{j}) + 10\,\mathbf{i} \times 30\,\mathbf{j}$$

$$15a\,\mathbf{k} = -50\,\mathbf{k} - 30\,\mathbf{k} + 300\,\mathbf{k}$$

$$\rightarrow a = 14.7\ \text{m}$$

Example 5.3

For the force-couple system applied to the structural beam below, calculate a single resultant force and locate this force relative to the wall.

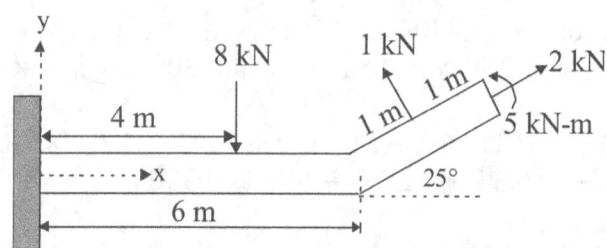

Solution:

$$\mathbf{R} = -8{,}000\,\mathbf{j} + 1{,}000\,(-\cos 65^\circ\,\mathbf{i} + \sin 65^\circ\,\mathbf{j}) + 2{,}000\,(\cos 25^\circ\,\mathbf{i} + \sin 25^\circ\,\mathbf{j})$$

$$\mathbf{R} = -8{,}000\,\mathbf{j} - 422.6\,\mathbf{i} + 906.3\,\mathbf{j} + 1{,}812.6\,\mathbf{i} + 845.2\,\mathbf{j}$$

$$\rightarrow \mathbf{R} = 1{,}390.0\,\mathbf{i} - 6{,}248.5\,\mathbf{j}\ \text{N}$$

The corresponding moment at the wall created by the applied forces is

$$\mathbf{M} = 4\,\mathbf{i} \times (-8{,}000\,\mathbf{j}) + [(6 + \cos 25°)\,\mathbf{i} + \sin 25°\,\mathbf{j}] \times (-422.6\,\mathbf{i} + 906.3\,\mathbf{j})$$
$$+ 6\,\mathbf{i} \times (1{,}812.6\,\mathbf{i} + 845.2\,\mathbf{j}) + 5{,}000\,\mathbf{k}$$

$$\mathbf{M} = -32{,}000\,\mathbf{k} + 6{,}259.2\,\mathbf{k} + 178.6\,\mathbf{k} + 5{,}071.2\,\mathbf{k} + 5{,}000\,\mathbf{k}$$

$$\rightarrow \mathbf{M} = -15{,}491.0\,\mathbf{k}\ \text{N-m}$$

Locating the line-of-action of $\mathbf{R}$ relative to point A using vector $\mathbf{a}$ shows

$$\mathbf{a} \times \mathbf{R} = \mathbf{M}$$

$$a\,\mathbf{i} \times (1{,}390.0\,\mathbf{i} - 6{,}248.5\,\mathbf{j}) = -15{,}491.0\,\mathbf{k}$$

$$-6{,}248.5\,a\,\mathbf{k} = -15{,}491.0\,\mathbf{k}$$

$$\rightarrow a = 2.48\ \text{m}$$

$$\rightarrow \mathbf{a} = 2.48\,\mathbf{i}\ \text{m}$$

Problems

5.1 The two-force system below creates a moment at point O equal to 750 $\mathbf{k}$ N-m. Write vector representations for the resultant force $\mathbf{R}$ (through point O) and force $\mathbf{F}$.
(Ans. $\mathbf{R} = 1{,}166.7\,\mathbf{j}$ N, $\mathbf{F} = 1{,}366.7\,\mathbf{j}$ N)

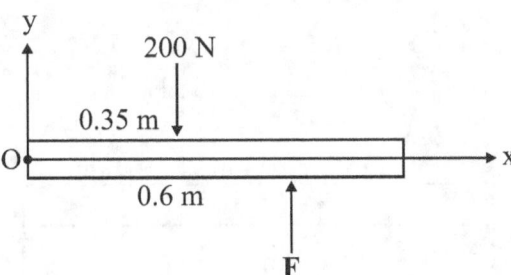

5.2 In the figure above, if F = 310 N, replace the two applied forces with a resultant force $\mathbf{R}$ and a moment $\mathbf{M}$ at point O.
(Ans. $\mathbf{R} = 110\,\mathbf{j}$ N, $\mathbf{M}_0 = 116\,\mathbf{k}$ N-m)

5.3 For the case of F = 175 lb in the figure below, calculate the resultant force and moment about point O of the system of forces shown.
(Ans. $\mathbf{R} = -301.7\,\mathbf{i} + 200.6\,\mathbf{j}$ lb, $\mathbf{M_O} = 1{,}783.2\,\mathbf{k}$ lb-in)

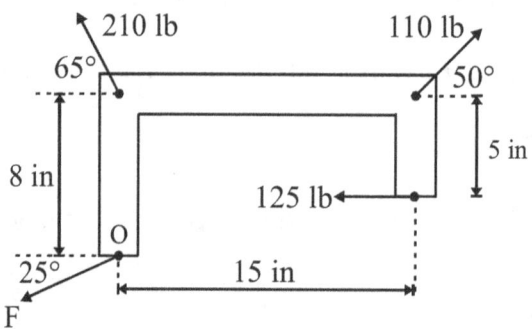

5.4 For the system of forces and the moment applied to the beam below, determine a single resultant force and locate this force relative to point O.
(Ans. $\mathbf{R} = -365\,\mathbf{j}$ N, $\mathbf{a} = 13.9\,\mathbf{i}$ m)

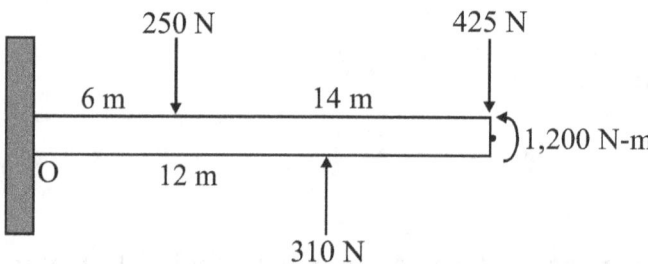

5.5 Calculate the resultant force of the force-couple system acting on the beam and locate this force relative to the wall.
(Ans. $\mathbf{R} = 2\,\mathbf{j}$ kN, $\mathbf{a} = 6.5\,\mathbf{i}$ m)

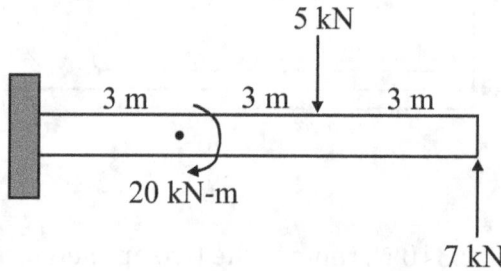

5.6 In the figure below, consider F = 195 lb and replace the system of forces with a single resultant force and locate this resultant relative to point O.
(Ans. **R** = −299.9 **i** + 166.4 **j** lb, **a** = 6.0 **i** in)

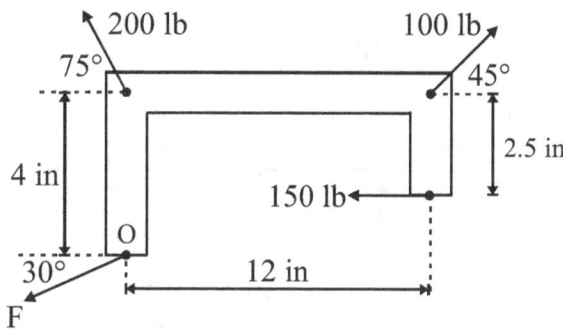

5.7 For the case of F = 140 lb in the figure above, compute the resultant force and moment about point O of the system of forces shown.
(Ans. **R** = −252.3 **i** + 193.9 **j** lb, **M**$_O$ = 997.7 **k** lb-in)

5.8 For the pulley system shown below, the resultant moment about point O is zero. Determine the tension T in the upper cable, the magnitude of the resultant force **R** acting through O, and the angle **R** makes with the positive x-axis for the case where F = 325 N.
(Ans. T = 390 N, R = 254.1 N, θ = 93.7° CCW from the x-axis)

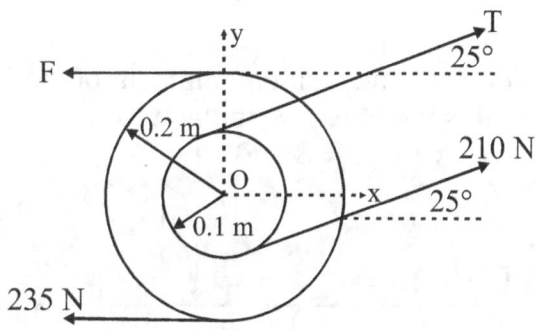

5.9 For the pulley system shown above, T = 325 N and the resultant moment about point O is zero. Calculate F, the magnitude of the resultant force **R** acting through O, and the angle **R** makes with the positive x-axis.
(Ans. F = 392.5 N, R = 340.2 N, θ = 66.0°)

5.10 The two-force system below creates a moment at point O equal to 600 **k** N-m. Determine representations for the resultant force **R** (through point O) and force **F**. (Ans. **R** = 728.6 j N, **F** = 1,028.6 j N)

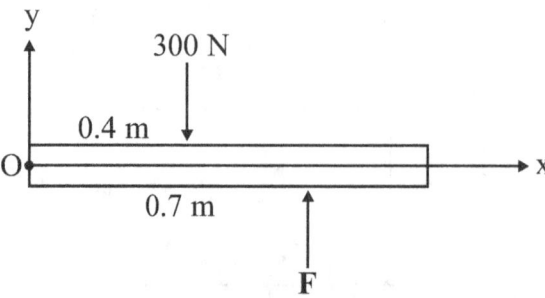

5.11 For the system of forces and the moment applied to the beam below, find a single resultant force and locate this force relative to point O. (Ans. **R** = −300 j N, **a** = 4.67 **i** m)

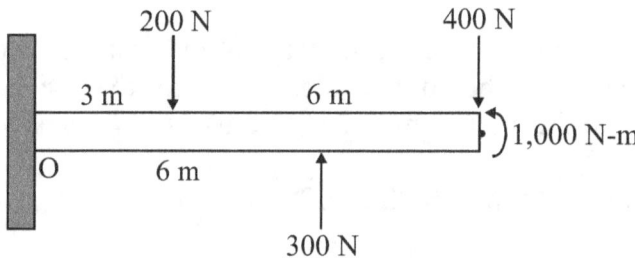

5.12 Determine the resultant force and resultant moment of the force-couple system acting on the beam and locate this force relative to the wall. (Ans. **R** = 2 j kN, **M**$_O$ = 6 **k** kN-m, **a** = 3 **i** m)

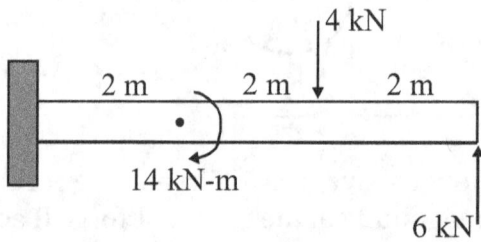

5.13 For the pulley system shown below, the resultant moment about point O is zero. Calculate the tension T in the upper cable, the magnitude of the resultant force **R** acting through O, and the angle **R** makes with the positive x axis for the case where F = 295 N.
(Ans. T = 290.7 N, R = 273.2 N, θ = 116.1° CCW from the x-axis)

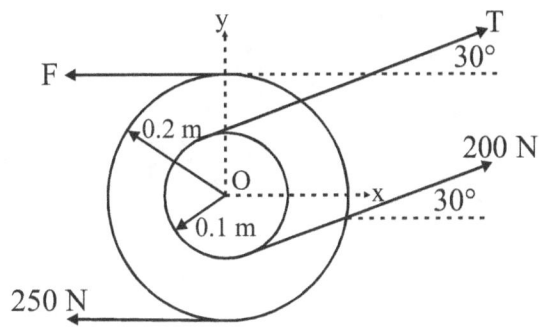

5.14 In Problem 5.10, if F = 700 N, replace the two forces with a resultant force **R** and a moment **M** at point O.
(Ans. **R** = 400 **j** N, **M**$_O$ = 370 **k** N-m)

5.15 Solve Problem 5.13 for F, R, and θ in the case of T = 325 N and F is unknown.
(Ans. F = 312.5 N, R = 283.8 N, θ = 112.3°)

(This page was intentionally left blank.)

MODULE 6: Equilibrium of Coplanar Force Systems

When a body is in equilibrium, the resultant of all forces and all moments acting on the body are equal to zero. Specific equilibrium conditions for two-dimensional problems are specified below.

Conditions of Equilibrium

Equilibrium of a body can be ensured by satisfying any of the following three sets of conditions:

I. Summing the forces in both directions to zero and summing the moments about <u>any</u> point A, on or off the body, to zero.

$$\sum F_x = 0 \quad , \quad \underline{\text{and}} \quad \sum F_y = 0 \quad , \quad \underline{\text{and}} \quad \sum M_A = 0$$

II. Summing the forces in one direction to zero and summing the moments about <u>any</u> two points, on the body to zero. Those points, A and B, aren't on the same line perpendicular to the direction of the summed forces.

$$\sum F_x = 0 \quad \underline{\text{or}} \quad \sum F_y = 0 \quad , \quad \underline{\text{and}} \quad \sum M_A = 0 \quad \underline{\text{and}} \quad \sum M_B = 0$$

III. Summing the moments about <u>any</u> three points on the body to zero, where those points, A, B, and C aren't on a straight line.

$$\sum M_A = 0 \quad , \quad \underline{\text{and}} \quad \sum M_B = 0 \quad , \quad \underline{\text{and}} \quad \sum M_C = 0$$

Each set of conditions can be used to solve for three unknowns in any problem. Fewer than three unknowns in a problem require fewer than three equations. Satisfying any of the three conditions above will guarantee that equilibrium exists in the problem being addressed.

Free-Body Diagrams

A free-body diagram (FBD) is a sketch of a body showing all the external forces and moments acting upon it. Drawing the free-body diagram is the single most important step in the solution of mechanics problems and is essential to correctly analyzing and solving problems.

External forces and moments to be shown on free-body diagrams include those applied from external loading conditions as well as those supplied by any supports or constraints attached to the body. Supports or constraints can take the form of cables or ropes, smooth or rough surfaces, roller supports, pinned connections which allow rotation, pinned connections which do not allow rotation, built-in or fixed supports, or free-sliding guides. These supports and constraints and their associated free-body diagrams are provided in Table 6.1 below.

Before attempting to solve equilibrium problems, some additional definitions are important in understanding possible solution methods. These definitions include statically determinate structures, statically indeterminate structures, and redundant supports.

'Statically determinate' structures are those that are supported by the minimum number of constraints necessary to maintain equilibrium. Problems involving these types of structures are solvable by standard solution methods.

'Statically indeterminate' structures are those that possess more external supports or constraints than are necessary to maintain equilibrium. Problems involving these types of structures are not solvable by standard solution methods.

'Redundant supports' are supports that can be removed without destroying the equilibrium conditions of a body. The existence of redundant supports in a structure usually makes that structure statically indeterminate.

Table 6.1 Two-Dimensional Supports

<u>Condition</u>   <u>Free-Body Diagram</u>

1. Flexible cable or rope.

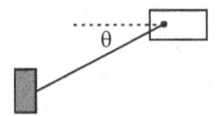   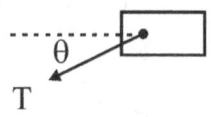

2. Smooth surface.

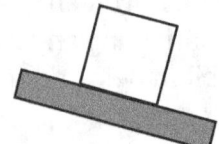   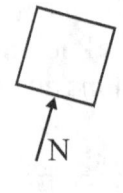

3. Rough surface.

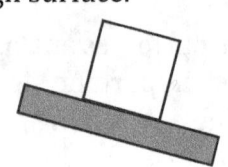   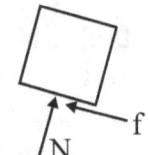

Table 6.1 Two-Dimensional Supports (cont.)

Condition   Free-Body Diagram

4. Simple and roller support.

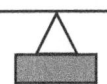 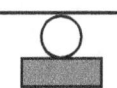

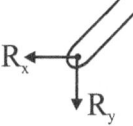

5. Pinned connection (free to rotate).

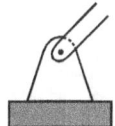

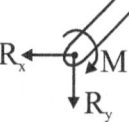

6. Pinned connection (no rotation).

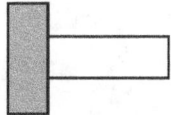

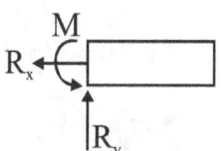

7. Built-in or fixed support.

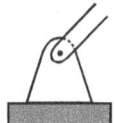

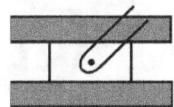

8. Free-sliding guide.

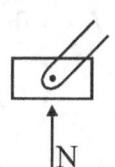

Example 6.1

For the beam shown below, calculate the reaction forces at supports A and B.

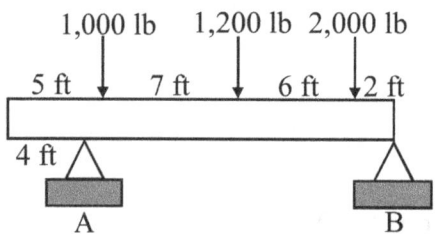

Solution:

The free-body diagram for this beam is

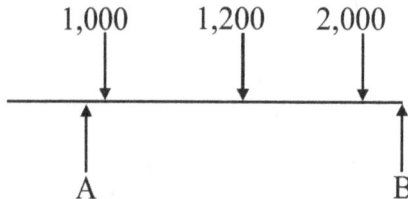

There are only two unknowns in this problem, the reaction forces in the upward direction at supports A and B. Therefore, the equilibrium of this beam can be shown by summing the forces in the vertical, or y-direction to zero and by summing the moments about any point on the beam to zero. There are no x-direction forces involved in this problem. Forces are summed in the y-direction as

$$\sum F_y = 0$$

$$A - 1,000 - 1,200 - 2,000 + B = 0$$

$$\rightarrow \ A + B = 4,200 \text{ lb}$$

It is often helpful to sum the moments about a point that will eliminate one of the unknowns from the moment equation. In this problem, summing the moments about either point A or B results in an equation having only one unknown.

$$\sum M_A = 0$$

$$\mathbf{i} \times (-1,000\,\mathbf{j}) + 8\,\mathbf{i} \times (-1,200\,\mathbf{j}) + 14\,\mathbf{i} \times (-2,000\,\mathbf{j}) + 16\,\mathbf{i} \times B\,\mathbf{j} = 0$$

$$-38,600\,\mathbf{k} + 16B\,\mathbf{k} = 0$$

→ B = 2,412.5 lb

→ A = 1,787.5 lb

Example 6.2

In the figure below, block A weighs 300 N and is connected to rod OB by a cable. If all surfaces are smooth, determine the value of force P applied at point B on the rod to maintain equilibrium.

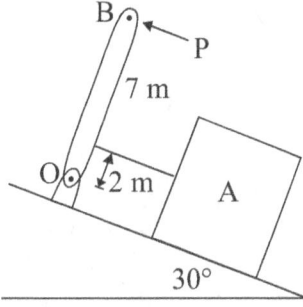

Solution:

In drawing the free-body diagrams for block A and rod OB, the coordinate system is chosen such that the x-axis points up the inclined plane and the y-axis is perpendicular to the plane.

Block A: Rod OB:

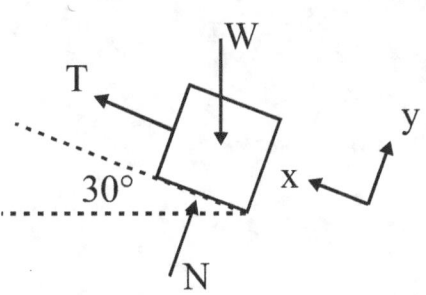

For block A, summing the forces in the x-direction gives

$$\Sigma F_x = 0$$

$$T - W \sin 30° = 0$$

$$T = W \sin 30° = 300(0.500)$$

$\rightarrow T = 150\ N$

$\rightarrow T = 150\ \mathbf{i}\ N$

For rod OB, if the moments are summed about point O an equation having only one unknown will be obtained.

$\sum \mathbf{M_O} = 0$

$9\ \mathbf{j} \times P\ \mathbf{i} + (2\ \mathbf{j}) \times (-150\ \mathbf{i}) = 0$

$-9P\ \mathbf{k} + 300\ \mathbf{k} = 0$

$\rightarrow P = 33.3\ N$

Therefore, a force of P = 33.3 N will be required to maintain equilibrium. If P has a value less than 33.3 N, the block will slide down the plane and the rod will rotate clockwise. If P has a value greater than 33.3 N, the rod will rotate counterclockwise and pull the block up the plane.

Example 6.3

For the structure below, find the tension in cable BC and the angle θ if the resultant of the applied forces at B is zero. Use a standard Cartesian coordinate system.

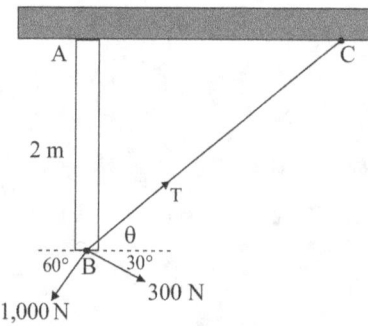

Solution:

Summing the applied forces at point B shows

$\mathbf{R} = \sum \mathbf{F} = -\ 1{,}000\ \cos 60°\ \mathbf{i} - 1{,}000\ \sin 60°\ \mathbf{j} + 300\ \cos 30°\ \mathbf{i} - 300\ \sin 30°\ \mathbf{j}$
$+ T\ \cos \theta\ \mathbf{i} + T\ \sin \theta\ \mathbf{j} = 0$

$\rightarrow\ -500.0\ \mathbf{i} - 866.0\ \mathbf{j} + 259.8\ \mathbf{i} - 150.0\ \mathbf{j} + T\ \cos \theta\ \mathbf{i} + T\ \sin \theta\ \mathbf{j} = 0$

Separating this equation into vector components gives

i direction: $-240.2 + T \cos \theta = 0$

j direction: $-1,016.0 + T \sin \theta = 0$

Solving these equations simultaneously for T and θ gives

$T = 1,044.0$ N

$\theta = 76.7°$

Problems

6.1 In Example 6.3, determine the reaction forces and moment at A for the case if
T = 400 N and θ = 40°. Consider the rod AB to have a mass of 30 kg.
(Ans. $A_x = -66.2$ N, $A_y = 1,053.2$ N, **M** $= -132.4$ **k** N-m)

6.2 The 17 ft long ladder shown in the figure below weighs 75 lb. If all surfaces are smooth,
calculate the force P necessary for the ladder not to slide.
(Ans. P = 23.0 lb)

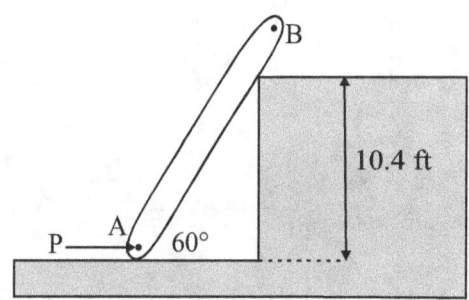

6.3 The ladder shown in the figure above is 15 ft long and it takes a force of P = 13. 5 lb to
keep the ladder from sliding. Compute the weight of the ladder if all surfaces are
considered to be smooth.
(Ans. W = 49.9 lb)

6.4 Calculate the value of P necessary to pull the wheel below over the step shown below. The wheel weighs 150 lb and has a radius of 15 in.
(Ans. **P** = 200.0 **i** lb)

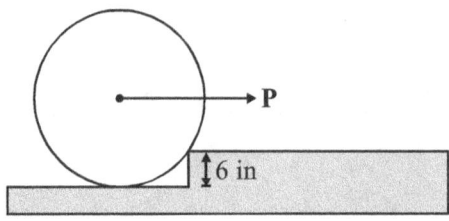

6.5 If the force necessary to begin pulling the 100 lb wheel over the step shown above is 124.9 lb, determine the diameter of the wheel.
(Ans. d = 32.0 in)

6.6 Calculate the tension in cable BC below if the W = 750 N. Assume all surfaces are smooth.
(Ans. T_{BC} = 750.0 N)

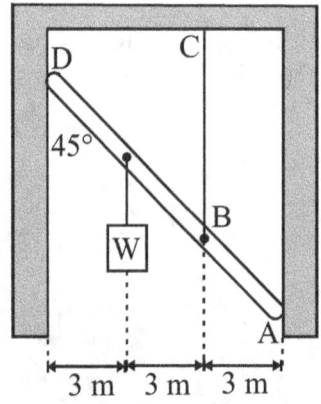

6.7 Calculate the weight of the block and the tension in cable BC above, if the reaction forces at A and D each equal 227 N.
(Ans. W 681 N, T_{BC} = 681.0 N)

6.8 For the cable system below, determine the tension in cables AC and BC.
(Ans. T_{AC} = 408.3 lb, T_{BC} = 644.0 lb)

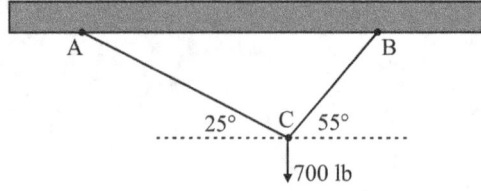

6.9 Compute the reaction force at point B for the structure shown below. Consider the pin at A to be free to rotate.
(Ans. B = 2,205.9 N)

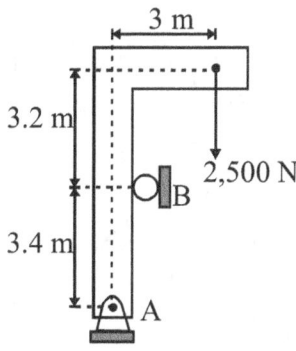

6.10 In the figure below calculate the value of P necessary to pull the wheel over the step. The wheel weighs 150 lb and has a radius of 15 in.
(Ans. $P = 284.3$ i lb)

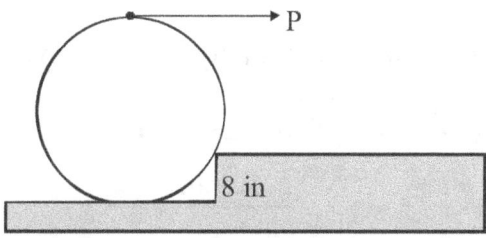

6.11 Find the weight of the wheel above if its diameter is 32 in and P = 173.2 lb.
(Ans. W = 100.0 lb)

6.12 For the cable system below, determine the tension in cables AC and BC.
(Ans. $T_{AC} = 261$ lb, $T_{BC} = 351.8$ lb)

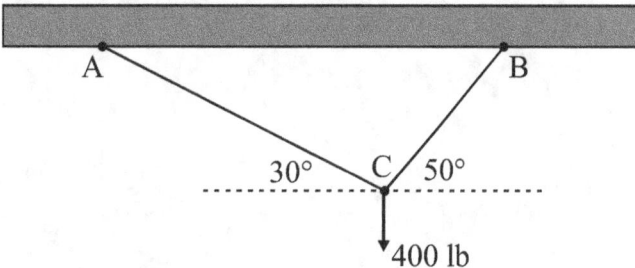

6.13 For a weight of 500 N in the figure below, find the reaction forces at A and D. All surfaces are smooth.

(Ans. F_A = 166.7 N, F_D = 166.7 N)

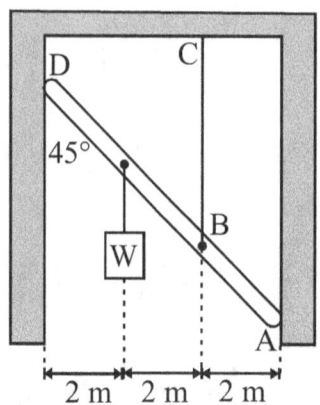

6.14 In Example 6.2, if P = 100 N, find the weight of the block.

(Ans. W = 900.0 N)

6.15 Compute the reaction forces at points A and B for the given structure, where the pin is free to rotate.

(Ans. A_x = 2,000.0 N, A_y = 2,000.0 N, B = 2,000.0 N)

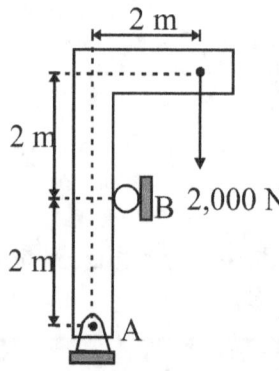

MODULE 7: Equilibrium of Coplanar Force Systems (cont.)

This module presents additional material on the equilibrium of coplanar force systems.

<u>Two-Force Members</u>

The analysis of many structures can be simplified by recognizing the existence of two-force members. A two-force member is one in which forces are only applied at the ends of the member and those two forces are equal, opposite, and collinear. Here the force transmitted through the member must act along an axis which connects the end points. This situation is depicted in Figure 7.1.

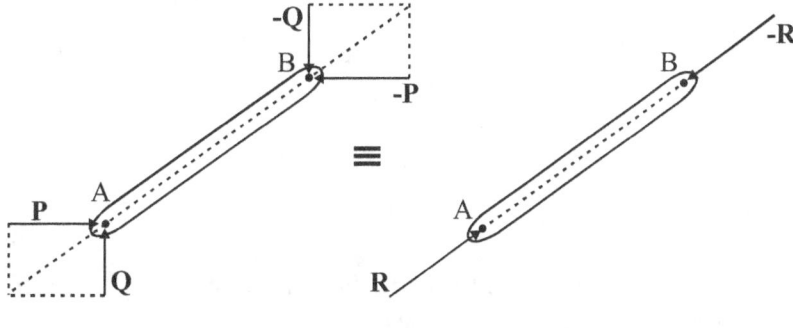

Figure 7.1

Here link AB is a two-force member and the resultant force **R** is directed along the axis AB. Therefore, it holds that

$$\mathbf{R} = \mathbf{P} + \mathbf{Q}$$

$$-\mathbf{R} = -\mathbf{P} - \mathbf{Q}$$

In solving problems, the identification of two-force members that exist in the structure is important, as it will simplify the analysis.

<u>Example 7.1</u>

For the linkage below, determine the force within links AB and BC.

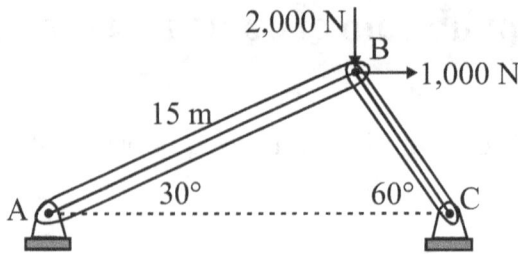

Solution:

Since links AB and BC both have loads applied only at the ends, they are each two-force members. Therefore, the forces transmitted through these links F_{AB} must be directed along their longitudinal axes. The free-body diagrams for these links are

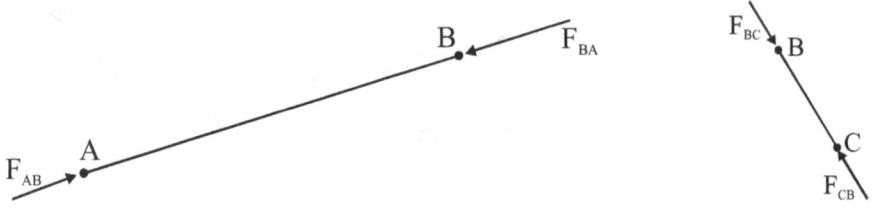

Both these links are shown to be in compression. Drawing a free-body diagram of the pin B connecting links AB and BC shows

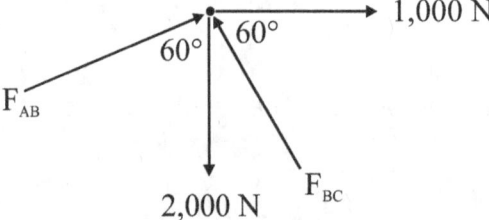

Summing the forces in both the x- and y- directions on the pin shows

$$\Sigma F_x = 0$$

$$F_{AB} \cos 30° - F_{CB} \cos 60° + 1{,}000 = 0$$

$$0.866F_{AB} - 0.500F_{CB} + 1{,}000 = 0$$

$$\Sigma F_y = 0$$

$$F_{AB} \sin 30° - F_{CB} \sin 60° - 2{,}000 = 0$$

$$0.500F_{AB} + 0.866F_{CB} - 2{,}000 = 0$$

Solving the two force equations simultaneously for F_{AB} and F_{CB} gives

$F_{AB} = 134.2$ N

$F_{CB} = 2,232.0$ N

Example 7.2

In the pinned linkage shown below, calculate the reaction forces at points A and C.

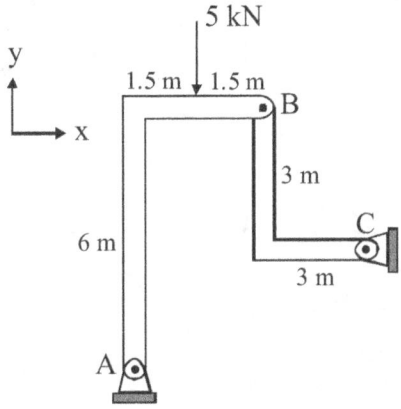

Solution:

Since there are 4 unknown forces to be determined, A_x, A_y, C_x, and C_y, and only three equilibrium equations, the structure must be broken-up into two separate parts, link AB and link BC. Also, since forces are only applied at the endpoints of link BC, BC is a two-force member. Therefore, the force carried by BC must be directed along the line connecting points B and C. The free-body diagrams for each part can be drawn as

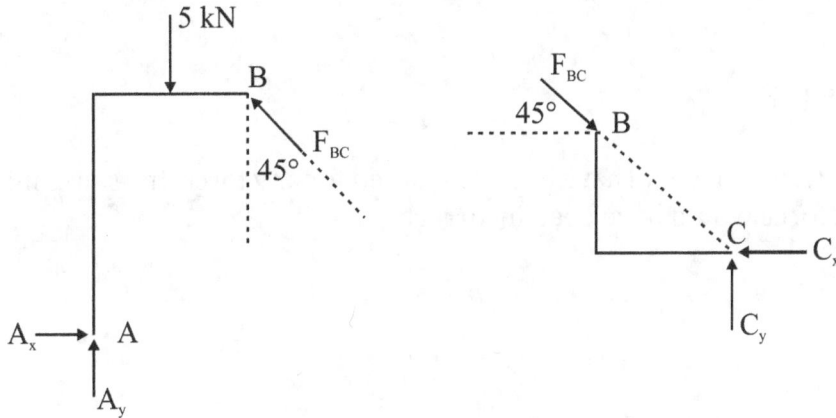

Link AB:

$$\sum F_x = 0$$

$$A_x - F_{BC} \cos 45° = 0$$

$$A_x - 0.707 F_{BC} = 0$$

$$\sum F_y = 0$$

$$A_x + F_{BC} \sin 45° - 5 = 0$$

$$A_x = 5 - 0.707 F_{BC}$$

$$\sum \mathbf{M_B} = 0$$

$$\left(-1.5\,\mathbf{i} \times (-5\,\mathbf{j})\right) + \left(-6\,\mathbf{j} \times A_x\,\mathbf{i}\right) + \left(-3\,\mathbf{i} \times A_y\,\mathbf{j}\right) = 0$$

$$7.5\,\mathbf{k} + 6A_x\,\mathbf{k} - 3A_y\,\mathbf{k} = 0$$

$$\rightarrow A_y = 2A_x + 2.5$$

Link BC:

$$\sum F_x = 0$$

$$F_{BC} \cos 45° - C_x = 0$$

$$C_x = 0.707 F_{BC}$$

$$\sum F_y = 0$$

$$F_{BC} \sin 45° + C_y = 0$$

$$C_y = 0.707 F_{BC}$$

Solving the five equilibrium equations above gives

$A_x = 0.83$ kN

$A_y = 4.17$ kN

$F_{BC} = 1.18$ kN

$C_x = 0.83$ kN

$C_y = 0.83$ kN

If a negative sign would have been calculated for any force, the sense initially assumed for that force would have been incorrect.

Problems

7.1 The beam below weighs 250 N and is pinned at point O. Determine the tension in cable AB and the reaction forces at O for the loading shown.
(Ans. T_{AB} = 877.1 N, O_x = 824.2 N, O_y = 125.0 N)

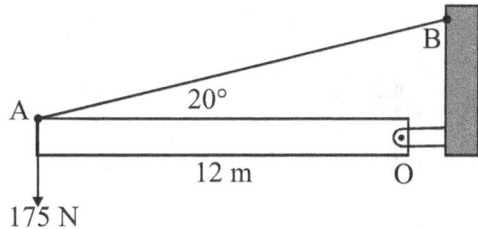

7.2 For the structure loaded as shown below, calculate the tension in cable AB and the reaction forces at pin C.
(Ans. T_{AB} = 6,743.7 lb, C_x = 6,111.8 lb, C_y = 1,250.0 lb)

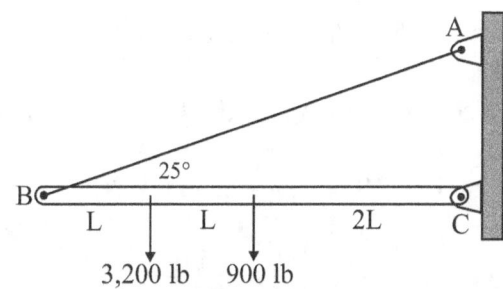

7.3 For the structure below, determine the reaction forces at C and the tension in cable AD if P = 65 N.
(Ans. C_x = 65.0 N, C_y = 65.0 N, T_{AB} = 65.0 N)

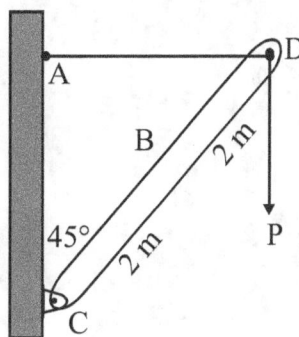

7.4 In the figure above, find the reaction forces at C and the tension in cable AB if P = 50 N and the weight of member CD is 150 N acting at its midpoint.
(Ans. T_{AD} = 125.0 N, C_x = 125.0 N, C_y = 200.0 N)

7.5 In the structure in Problem 7.3, if P = 125 N, compute the weight of member CD and the reaction forces at point C if the tension in cable AD is 230 N.
(Ans. W = 210.0 N, C_x = 230.0 N, C_y = 335.0 N)

7.6 Calculate the mass of block M, which will create a force of 400 N in member BC below.
(Ans. M = 16.7 kg)

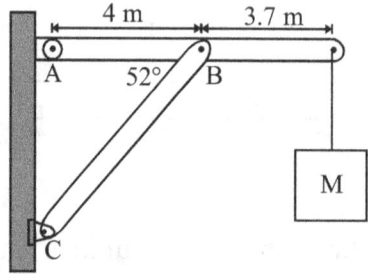

7.7 A mass of 14 kg is supported by the structure above. Determine the force in member BC and the reaction forces at pin A.
(Ans. F_{BC} = 335.5 N, A_x = 206.7 N, A_y = −127.0 N)

7.8 Compute the reaction forces at point A and the tension in cable BC for the given structure.
(Ans. A_x = 1,662.5 N, A_y = 3,295.0 N, T_{BC} = 2,170.2 N)

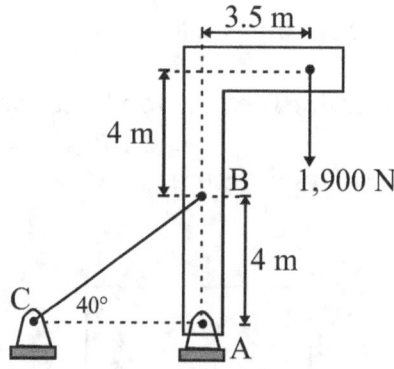

7.9 A mass of 10 kg is supported by the structure as shown. Calculate the force in member BC and the reaction forces at pin A.
(Ans. $F_{BC} = 245.6$ N, $A_x = 167.5$ N, $A_y = -81.7$ N)

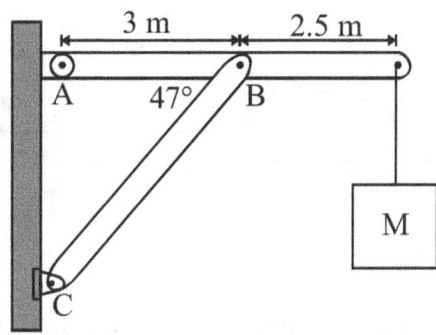

7.10 In the figure above, find the force in member BC if the mass M is 16.3 kg.
(Ans. $F_{BC} = 400.8$ N)

7.11 For the structure below, determine the load P and the reaction forces at C if the tension in cable AB is 100 N.
(Ans. P = 50.0 N, $C_x = 100.0$ N, $C_y = 50.0$ N)

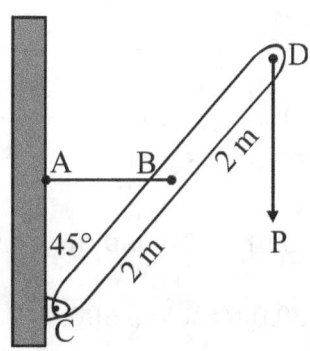

7.12 In Problem 7.11, find the weight of member CD if P = 65 N and $T_{AB} = 280$ N.
(Ans. W = 150.0 N)

7.13 The beam below weighs 200 N and is pinned at point O. Compute the tension in cable AB and the reaction forces at O for the loading shown.
(Ans. T_{AB} = 472.8 N, O_x = 428.4 N, O_y = 100.0 N)

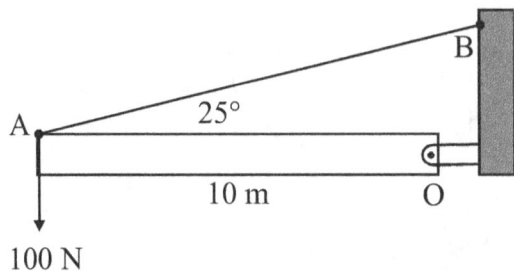

7.14 For the structure loaded as shown below, calculate the tension in cable AB and the reaction forces at pin C for the case where L = 8 ft.
(Ans. T_{AB} = 6,212.9 lb, C_x = 5,840.1 lb, C_y = 875.0 lb)

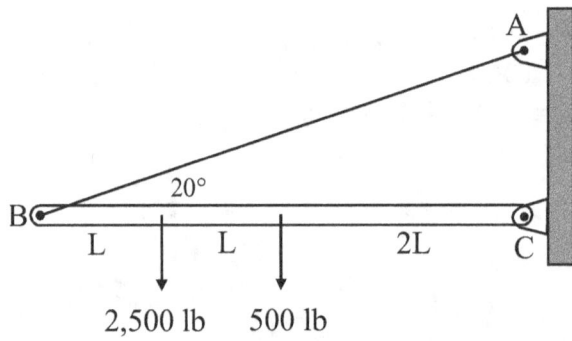

7.15 Determine the reaction forces at point A and the tension in cable BC for the given structure.
(Ans. T = 2,121.3 N, A_x = 1,500.0 N, A_y = 3,500.0 N)

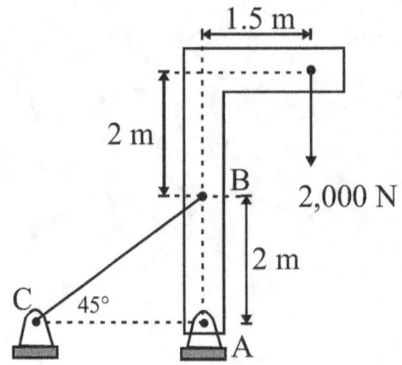

MODULE 8: Trusses

Some of the most important structural members used in engineering are trusses. Trusses are structures composed of individual rigid elements connected to form one or more triangular sections, with the individual elements being pinned together at joints. The rigid elements which form a truss lie in the same plane, therefore only coplanar force systems are involved. The weight of the members is neglected initially and all external forces are applied at the joints. All the elements composing a truss are two-force members and are therefore loaded in either tension or compression. If a truss is loaded in tension, the force within the member will pull on the joint, and if a truss is loaded in compression, the force within the member will push on the joint. Examples of simple trusses are shown in Figure 8.1.

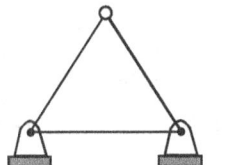 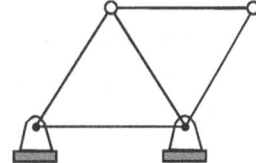 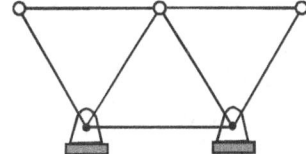

Figure 8.1

Since trusses are formed using triangular sections, it is often unnecessary to know the lengths of the members or the overall dimensions of the truss. In analyzing trusses, it is important to determine the axial forces in each member. There are two methods used in the analyses of trusses, the Method of Joints and the Method of Sections.

The Method of Joints considers free-body diagrams of each joint and solves two-dimensional force equilibrium equations to determine the force in each member at the joint. This method is useful in determining the forces that must be carried by the pinned connections. However, this method is limited to only being able to solve for two unknown forces at each joint.

The Method of Sections cuts the truss through certain members to form smaller sections of the truss. Two-dimensional equilibrium equations, as well as a moment equations about any point in the section are then solved to determine the forces being carried through the members that are cut. Therefore, this method is able to solve for three unknown forces in each section and is more useful in determining the internal forces of selected members.

Due to its usefulness, only The Method of Sections will be addressed in this text. The general procedure and examples of the use of the Method of Sections are provided below.

Method of Sections

The general procedure for the application of the Method of Sections is:

1. Draw a free-body diagram of the complete structure.

2. Use force and moment equilibrium equations to determine the support reaction forces.

3. Identify the members whose internal forces need to be determined.

4. Cut through the members of the truss to be analyzed up to a maximum of three members creating smaller sections of the truss.

5. Draw a free-body diagram of one portion of the sectioned truss, showing all externally applied loads, support loads, and the internal loads carried by the members cut. Assume either a tension or compression for the unknown loads in these members.

6. Using two-dimensional equilibrium equations for the section being considered, calculate the loads carried by the cut members. If a force is calculated to be positive, the direction assumed (tension or compression) was correct. If a force is calculated to be negative, the direction assumed was incorrect.

Example 8.1

For the truss shown, determine the forces in members BC, FC, and FG.

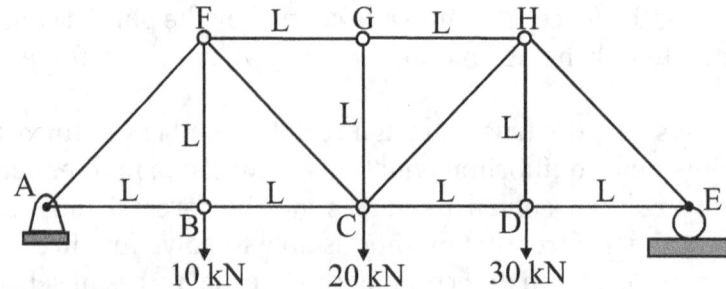

Solution:

Drawing a free-body diagram of the complete truss shows

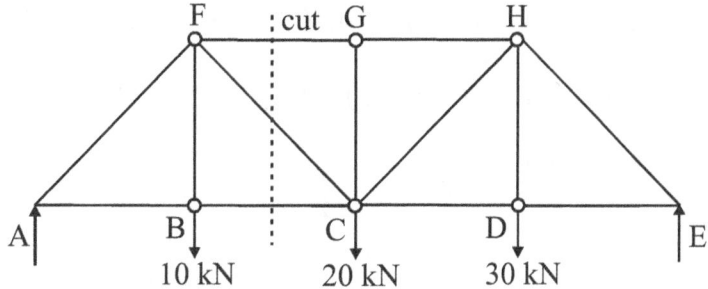

The reaction forces at the supports are found by equilibrium equations to be

A = 25.0 kN

E = 35.0 kN

To determine the forces in members BC, FC, and FG, the truss must be cut through those three members as shown above, creating two portions of the truss. Either of the two potions can be analyzed to find the forces in members BC, CF, and FG. Analyzing the left portion of the truss shows

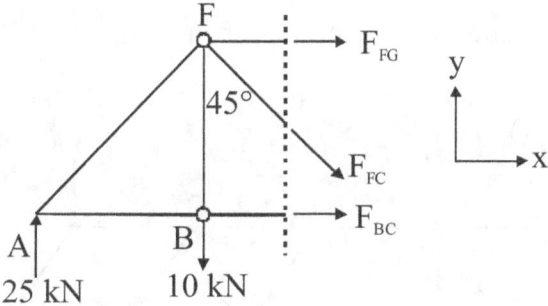

All the members are assumed to be in tension and the length of each horizontal and vertical member is assumed to be L. The unknown forces F_{BC}, F_{FC}, and F_{FG} can be determined by solving the equilibrium conditions as

$\Sigma F_y = 0$

$25.0 - 10.0 - F_{FC} \cos 45° = 0$

$\rightarrow F_{FC} = 21.2$ kN $= 21.2$ kN T (tension)

$$\sum M_A = 0$$

$$L\,\mathbf{i} \times (-10.0\,\mathbf{j}) + (L\,\mathbf{i} + L\,\mathbf{j}) \times F_{FG}\,\mathbf{i} + (L\,\mathbf{i} + L\,\mathbf{j}) \times (F_{FC}\sin 45°\,\mathbf{i} - F_{FC}\cos 45°\,\mathbf{j}) = 0$$

$$-10.0L\,\mathbf{k} - F_{FG}L\,\mathbf{k} - (0.707)F_{FC}L\,\mathbf{k} - (0.707)F_{FC}L\,\mathbf{k} = 0$$

$$-10.0 - F_{FG} - (0.707)(21.2) - (0.707)(21.2) = 0$$

$$\rightarrow F_{FG} = -40.0\text{ kN} = 40.0\text{ kN C (compression)}$$

$$\sum F_X = 0$$

$$F_{BC} + F_{FC}\cos 45° + F_{FG} = 0$$

$$F_{BC} + (21.2)(0.707) + (-40.0) = 0$$

$$\rightarrow F_{BC} = 25.0\text{ kN} = 25.0\text{ kN T}$$

From these results, members FC and BC will be in tension (T) and member FG will be in compression (C).

Example 8.2

For the truss shown below, determine the forces in members AB, BC, and CD. All members have a length equal to d.

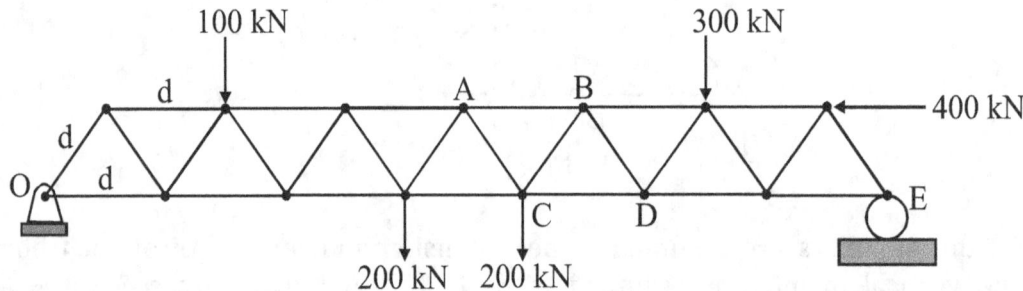

Solution:

Free-body diagram:

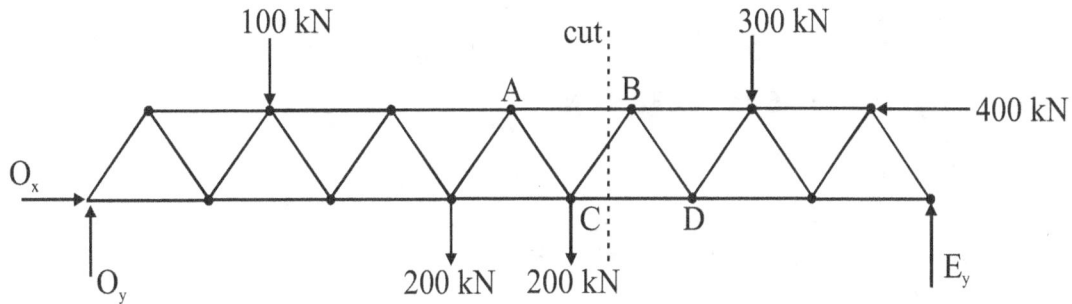

The reaction forces at the supports are

$O_x = 400.0$ kN

$O_y = 392.3$ kN

$E_y = 407.7$ kN

Cutting the truss as shown will give the right portion as

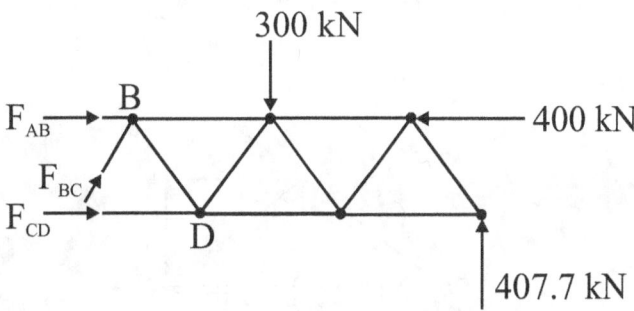

The three members in question are each assumed to be in compression. Solving for the unknown forces shows

$$\sum F_y = 0$$

$$407.7 - 300.0 + F_{BC} \sin 60° = 0$$

$$\rightarrow F_{BC} = -124.4 \text{ kN} = 124.4 \text{ kN T}$$

$$\sum M_B = 0$$

$$d\,\mathbf{i} \times (-300.0\,\mathbf{j}) + 2.5d\,\mathbf{i} \times (407.7\,\mathbf{j}) + (-d\cos 30°)\,\mathbf{j} \times F_{CD}\,\mathbf{i} = 0$$

$$-300d\,\mathbf{k} + 1{,}019.3d\,\mathbf{k} + 0.866dF_{CD}\,\mathbf{k} = 0$$

$$\rightarrow F_{CD} = -830.6 \text{ kN} = 830.6 \text{ kN T}$$

$$\sum F_x = 0$$

$$F_{CD} + F_{AB} + F_{BC}\cos 60° - 400.0 = 0$$

$$-830.6 + F_{AB} + (-124.4)(0.500) - 400.0 = 0$$

$$\rightarrow F_{AB} = 1{,}292.8 \text{ kN} = 1{,}292.8 \text{ kN C}$$

This result indicates that member AB is in compression (C) and members BC and CD are in tension (T).

Problems

8.1 Calculate the forces in members AD, BC, and BD when P = 80 kN.
(Ans. F_{AD} = 96.2 kN T, F_{BC} = 0.0 kN, F_{BD} = 0.0 kN)

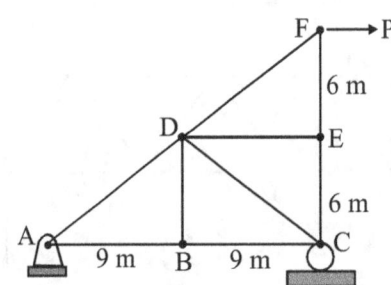

8.2 In Example 8.1, determine the forces in members CD, CH, and GH.
(Ans. F_{CD} = 35.0 kN T, F_{CH} = 7.1 kN T, F_{GH} = 40.0 kN C)

8.3 Compute the forces in members BC, CE, and EF for the condition of P = 850 N.
(Ans. F_{BC} = 2,243.0 N T, F_{CE} = 271.2 N C, F_{EF} = 4,556.9 N C)

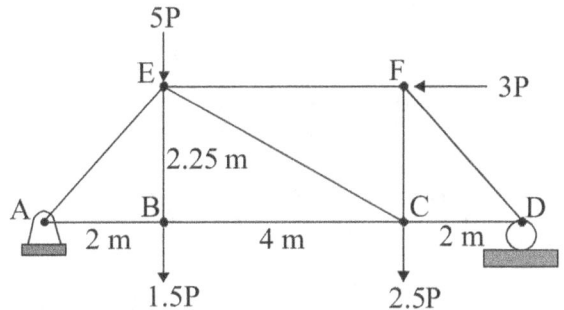

8.4 In the truss shown below, find the forces in members FG, BG, and BC.
(Ans. F_{FG} = 26.7 kN C, F_{BG} = 24.1 kN C, F_{BC} = 40.1 kN T)

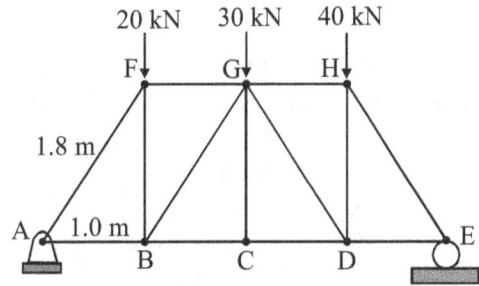

8.5 Calculate the forces in members CD, CG, and FG of the truss below if P = 1,500 lb.
(Ans. F_{CD} = 765.9 lb T, F_{CG} = 664.2 lb T, F_{FG} = 1,098.7 lb C)

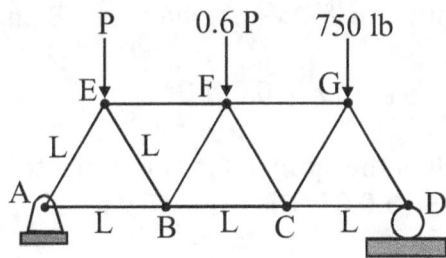

8.6 In the truss above, determine the value of P which would create a tension force of
1,000 lb in member FG.
(Ans. P = 1,317.2 lb)

8.7 Find the forces in members CD, CG, and FG in the truss below.
(Ans. F_{CD} = 19,052.6 N T, F_{CG} = 7,794.2 N C, F_{FG} = 17,500.0 N C)

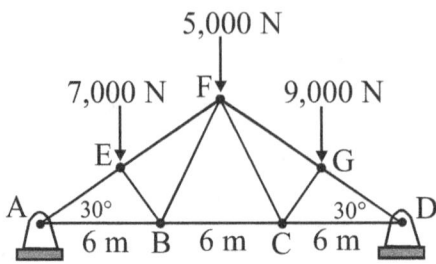

8.8 Calculate the forces in members BC, BF, and EF in the truss above.
(Ans. F_{BC} = 11,258.3 N T, F_{BF} = 6,062.2 N T, F_{EF} = 16,500.0 N C)

8.9 In Problem 8.5, determine the forces in members CD, CG, and FG if P = 1,000 lb.
(Ans. F_{CD} = 630.3 lb T, F_{CG} = 394.6 lb T, F_{FG} = 827.6 lb C)

8.10 In the truss below, find the forces in members CD, CG, and FG.
(Ans. F_{CD} = 17,754.0 N T, F_{CG} = 6,928.5 N C, F_{FG} = 16,500.2 N C)

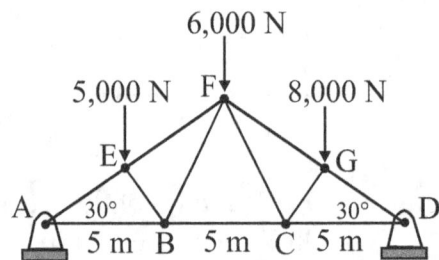

8.11 In the truss above, compute the forces in members AB and AE for the loading condition shown.
(Ans. F_{AB} = 15,155.4 N T, F_{AE} = 17,500.0 N C)

8.12 Solve Example 8.1 if each of the applied loads is equal to 30 kN.
(Ans. F_{FC} = 21.1 kN T, F_{FG} = 60.0 kN T, F_{BC} = 45.0 kN T)

8.13 Calculate the forces in members AD, BC, and BD below when P = 50 kN.
(Ans. F_{AD} = 62.5 kN T, F_{BC} = 0.0 kN, F_{BD} = 0.0 kN)

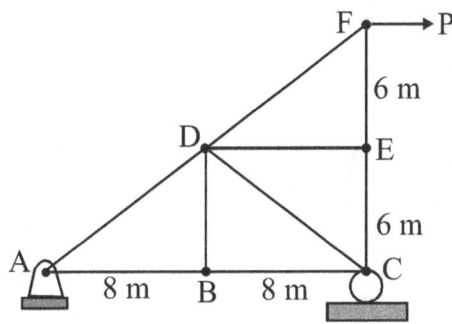

8.14 In the frame above, find the force carried by member DE for P = 75 kN.
(Ans. F_{DE} = 0.0)

8.15 In Problem 8.10, determine the forces in members BC, BF, and EF.
(Ans. F_{BC} = 10,825.3 N T, F_{BF} = 4,330.1 N T, F_{EF} = 15,000.0 N C)

(This page was intentionally left blank.)

MODULE 9: Friction

Friction forces result when two surfaces that possess some roughness, i.e., not smooth, move or attempt to move relative to each other. Two types of friction forces exist: static friction forces and kinetic friction forces. Static friction is the tangential force that opposes the sliding of one body relative to another. Kinetic friction is the tangential force between two bodies after motion begins. The magnitudes of these two types of friction forces are given as

$$f_s \leq \mu_s N$$

$$f_k = \mu_k N$$

where

$$\mu_s = \text{coefficient of static friction}$$

$$\mu_k = \text{coefficient of kinetic friction}$$

Since μ_k is usually smaller than μ_s, the magnitude of the kinetic friction force is less than the magnitude of the static friction force. The coefficients of friction are functions of the materials of the two bodies and the roughness of the surfaces. The relationship between static friction and kinetic friction can be seen in Figure 9.1, where F is applied to a block resting on a surface.

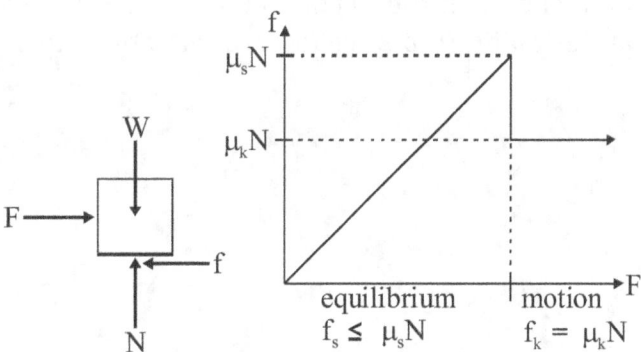

Figure 9.1

This figure shows that as the applied force increases from zero, the static friction force resisting the motion between the block and the surface increases during the equilibrium (or no-slipping) phase. When motion of the block is impending, i.e., motion ready to begin, the magnitude of the maximum static friction force is achieved and is equal to

$$f_{max} = \mu_s N$$

When motion (or slipping) of the block begins, the magnitude of the friction force drops to a value equal to the kinetic friction force, which will be constant as along as the block remains in motion.

Example 9.1

Consider a block of mass m positioned on a plane inclined at an angle θ. Determine the maximum value of θ for equilibrium.

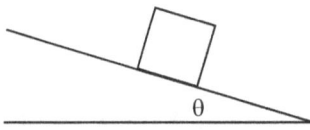

Solution:

Assuming the x-axis to be parallel to the inclined plane and the y-axis to be perpendicular to the inclined plane, the free-body diagram of the block will be

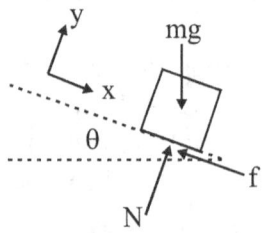

Since motion is impending at the maximum value of θ, the static friction force will be a maximum at this point. Summing the force equations to zero gives

$$\sum F_x = 0$$

$$W \sin \theta - f_{max} = 0$$

$$W \sin \theta = \mu_s N$$

$$\rightarrow N = \frac{W \sin \theta}{\mu_s}$$

$$\sum F_y = 0$$

$$-W \cos \theta + N = 0$$

$$\rightarrow \frac{W \sin \theta}{\mu_s} = W \cos \theta$$

$$\rightarrow \theta = \tan^{-1} \mu_s \quad \text{(Note: θ is only a function of } \mu_s \text{ and not the mass of the block.)}$$

Example 9.2

For the two-block system shown below, calculate the value of the force P for impending motion up the plane. Block A weighs 20 N and block B weighs 15 N. Assume that the pulley is frictionless and the coefficient of static friction between all other surfaces is 0.25.

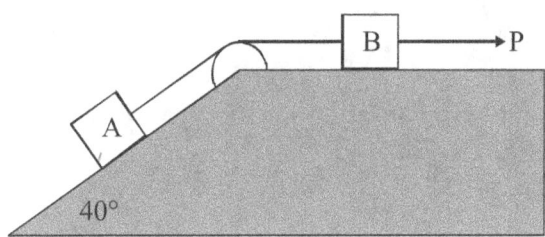

Solution:

Drawing the free-body diagrams for each block shows

Since motion is impending, the static friction forces will be maximized. Evaluating the equilibrium conditions for each bock gives

Block A:

$$\sum F_x = 0$$

$$T - f_A - W_A \sin \theta = 0$$

$$T - \mu_s N_A - W_A \sin \theta = 0$$

$$T = 0.25 N_A + 20 \sin 40°$$

$$\rightarrow T = 0.25 N_A + 12.86$$

Block B:

$$\sum F_x = 0$$

$$P - T - f_B = 0$$

$$P - T - \mu_s N_B = 0$$

$$\rightarrow P = T + 0.25 N_B$$

$$\Sigma F_y = 0 \qquad\qquad\qquad \Sigma F_y = 0$$

$$N_A - W_A \cos\theta = 0 \qquad\qquad N_B - W_B = 0$$

$$N_A = 20\cos 40° \qquad\qquad\qquad \rightarrow N_B = 15.0 \text{ N}$$

$$\rightarrow N_A = 15.3 \text{ N}$$

Substituting and solving the x-direction force equations for T and P gives

$$T = 0.25(15.3) + 12.86$$

$$\rightarrow T = 16.7 \text{ N}$$

$$P = 16.7 + 0.25(15.0)$$

$$\rightarrow P = 20.5 \text{ N}$$

Problems

9.1 Block A has a mass of 25 kg and block B has a mass of 80 kg. If P = 348.6 N, find the force in the cable attached at C. The coefficient of friction between all surfaces is 0.3. (Ans. T_{AC} = 80.2 N)

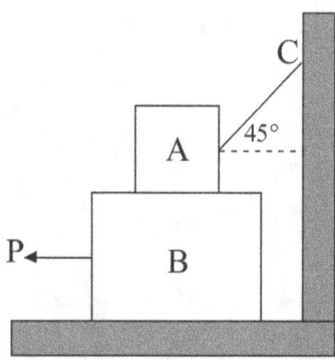

9.2 In the figure above Block A has a mass of 20 kg and block B has a mass of 60 kg. Determine the magnitude of force P in order to move block B. The coefficient of friction between all surfaces is 0.35. (Ans. P = 307.8 N)

9.3 The ladder shown in the figure below weighs 175 lb. Considering that $\mu_s = 0.25$ calculate the force P necessary for the ladder not to slide.
(Ans. P = 8.5 lb)

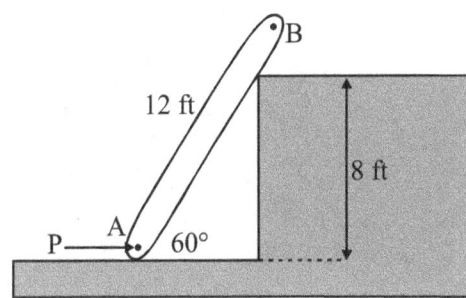

9.4 In the figure above, if P = 34.4 lb keeps the ladder from sliding, find the weight of the ladder. Assume that $\mu_s = 0.15$,
(Ans. W = 250 lb)

9.5 Determine the minimum value of P necessary to start motion in the two-block system shown. The coefficient of static friction between all surfaces is 0.2 and M = 15 kg.
(Ans. P = 201.6 N)

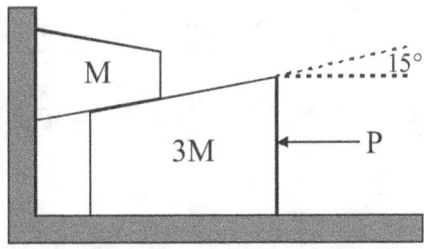

9.6 For the system shown, calculate the range of values of W to maintain equilibrium. Block A weighs 200 lb and the coefficient of friction between the blocks and the inclined surfaces is 0.38. Assume that the pulley is frictionless.
(Ans. 14.7 lb < W < 205.6 lb)

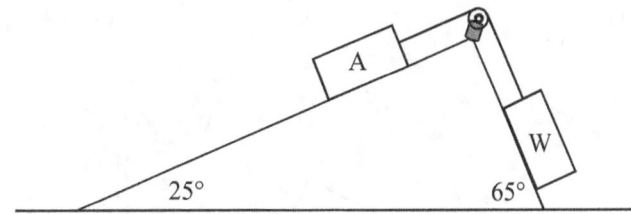

9.7 Block A weighs 125 N, block B weighs 175 N, and $\mu_s = 0.35$ in the system below. Compute the maximum value P can have before slipping occurs.
(Ans. P = 208.8 N)

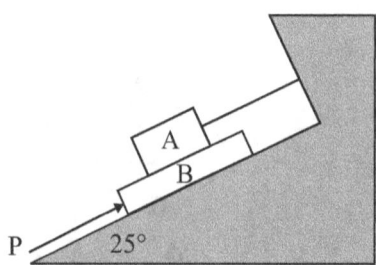

9.8 In the two-block system shown below, find the minimum value of P necessary to start motion. The coefficient of static friction between all surfaces is 0.2 and M = 15 kg.
(Ans. P = 69.7 N)

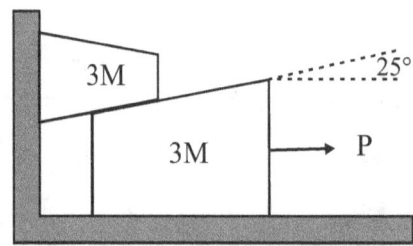

9.9 In Example 9.2, calculate the mass of block A for impending motion down the plane for the case of P = 10 N.
(Ans. $m_A = 3.1$kg)

9.10 Assuming the mass of block A is unknown in Example 9.2, determine its mass for impending motion down the plane for the case of P = 0.
(Ans. $m_A = 0.85$ kg)

9.11 In the figure below, block A weighs 125 N and block B weighs 175 N. If $\mu_s = 0.35$, find the maximum value P before slipping occurs.
(Ans. P = 60.9 N)

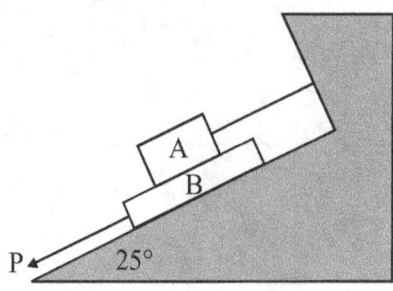

9.12 In the figure above, block A weighs 100 N and block B weighs 150 N. If the maximum value of P before slipping occurs is 31.7 N, determine the coefficient of friction between the surfaces.
(Ans. $\mu_s = 0.30$)

9.13 Calculate the minimum value of P necessary to start motion in the two-block system shown. The coefficient of static friction between all surfaces is 0.3 and M = 10 kg.
(Ans. P = 218.4 N)

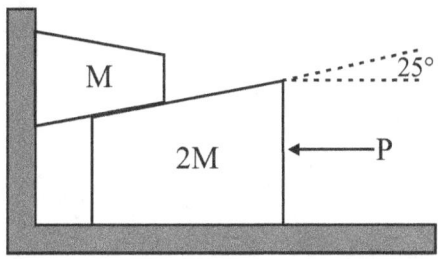

9.14 For the system shown, determine the range of values of W for block A to maintain equilibrium. The weight of block A is 125 lb and the coefficient of friction between the blocks and the inclined surfaces is 0.4. Assume that the pulley is frictionless.
(Ans. 18.0 lb < W < 158.9 lb)

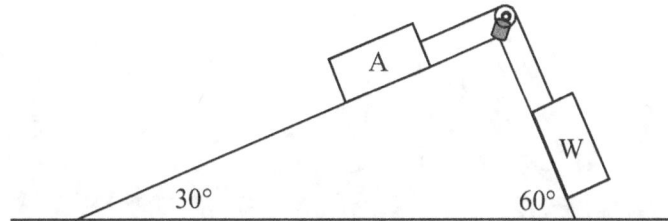

9.15 In Problem 9.12, assume that P acts up the plane rather than down the plane. Find the maximum value of P before slipping occurs.
(Ans. P = 158.6 N)

(This page was intentionally left blank.)

MODULE 10: Friction (cont.)

The previous module addressed friction problems in which motion was occurring or impending. In those cases, the friction forces can be easily calculated. If motion is occurring, the friction force is equal to the kinetic friction force, and if motion is impending, the friction force is equal to the maximum possible static friction force. Problems of a more challenging nature are ones in which the analysis must determine the magnitude of the friction force present and whether motion is occurring or impending. The following examples demonstrate those types of friction problems.

Example 10.1

In the two-block system shown below, block A has a mass of 4 kg and block B has a mass of 3 kg. The coefficient of static friction between blocks A and B is 0.35, and the coefficient of static friction between block B and the surface is 0.25. Determine the value of the applied force P for impending motion of block A and determine if block B is slipping.

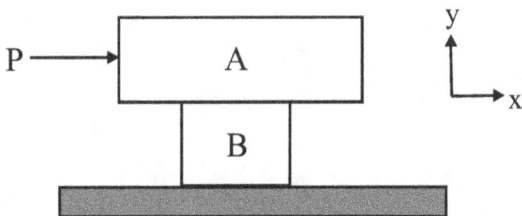

Solution:

Drawing free-body diagrams for each block shows

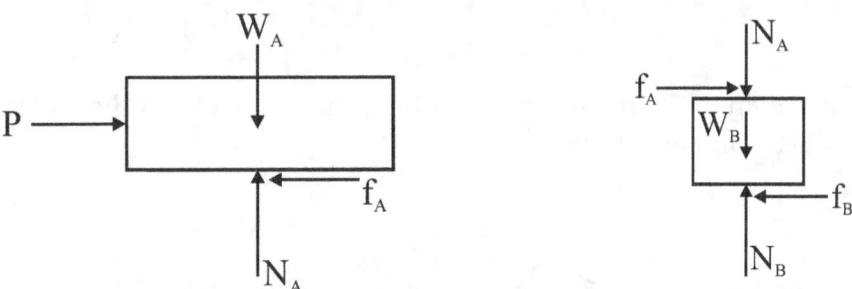

Since motion of A is impending, f_A will be a maximum. The equilibrium conditions for each block give

Block A: Block B:

$$\Sigma F_y = 0$$ $$\Sigma F_y = 0$$

$$N_A - m_A g = 0 \qquad\qquad N_B - N_A - m_B g = 0$$

$$N_A = m_A g \qquad\qquad N_B = N_A + m_B g$$

$$\rightarrow N_A = 4(9.81) = 39.2 \text{ N} \qquad\qquad \rightarrow N_B = 39.2 + 3(9.81) = 68.6 \text{ N}$$

$$\Sigma F_x = 0 \qquad\qquad \Sigma F_x = 0$$

$$P - f_A = 0 \qquad\qquad f_A - f_B = 0$$

$$P = \mu_A N_A \qquad\qquad f_B = f_A = \mu_A N_A$$

$$P = 0.35(39.2) \qquad\qquad f_B = 0.35(39.2)$$

$$\rightarrow P = 13.7 \text{ N} \qquad\qquad \rightarrow f_B = 13.7 \text{ N}$$

The actual friction force between block B and the surface must be compared to the maximum possible friction force, computed by

$$f_{B-max} = \mu_B N_B$$

$$F_{B-max} = 0.25(68.6)$$

$$\rightarrow f_{B-max} = 17.2 \text{ N}$$

Since the actual friction force is less than the maximum possible friction force, there will not be any slipping between block B and the surface.

Example 10.2

The block below has a weight of 75 lb. The coefficients of friction between the block and the inclined plane are $\mu_s = 0.30$ and $\mu_k = 0.25$.

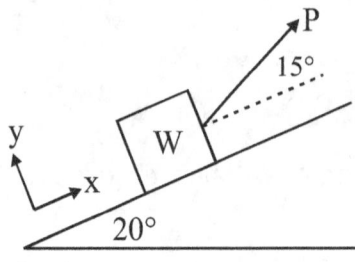

Calculate the friction force for the following cases: (a) $P = 0$, (b) $P = 35$ lb, (c) $P = 70$ lb, and (d) find P for impending motion of the block up the plane.

Solution:

The free-body diagram for the block is drawn below. Note that the direction of the friction force can be either up or down the plane, depending on the direction of motion or impending motion. It will be assumed that the friction force will be acting in the upward direction on the block. If the friction force is calculated to be negative, the sense will be opposite to the assumed direction.

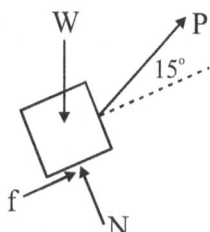

In analyzing problems of this type, it is often necessary to assume that either equilibrium exists, or that slipping occurs. Whichever assumption is made can then be proven to be valid or invalid based upon the results of the calculations. Therefore, in solving this problem it will be assumed that equilibrium conditions exist. Summing the forces acting on the block to zero gives

$\Sigma F_x = 0$

$P \cos 15° + f - W \sin 20° = 0$

$\rightarrow f = 25.65 - 0.966P$

$\Sigma F_y = 0$

$N + P \sin 15° - W \cos 20° = 0$

$\rightarrow N = 70.48 - 0.259P$

(a) For $P = 0$:

$f = 25.65 - 0.966(0) = 25.65 \text{ lb}$

$N = 70.48 - 0.259(0) = 70.48 \text{ lb}$

The maximum possible friction force will be

$f_{max} = \mu_s N = 0.30(70.48) = 21.1 \text{ lb}$

Since $f > f_{max}$, the original assumption of equilibrium is invalid. Therefore, the block will slip and the friction force will be equal to the kinetic friction force calculated by

$$f = \mu_k N = 0.25(70.48) = 17.6 \text{ lb}$$

(b) For $P = 35$ lb:

$$f = 25.65 - 0.966(35) = -8.2 \text{ lb} \quad \text{(acting downward on the block)}$$

$$N = 70.48 - 0.259(35) = 61.4 \text{ lb}$$

The maximum friction force is

$$f_{max} = \mu_s N = 0.30(61.4) = 18.4 \text{ lb}$$

Since $f < f_{max}$ in the case, the original assumption of equilibrium is valid.

(c) For $P = 70$ lb:

$$f = 25.65 - 0.966(70) = -42.0 \text{ lb}$$

$$N = 70.48 - 0.259(70) = 52.4 \text{ lb}$$

The negative sign on the friction force shows the direction is upward with a magnitude of 41.968 lb and the maximum friction force is

$$f_{max} = \mu_s N = 0.30(52.4) = 15.7 \text{ lb}$$

Since $f > f_{max}$, the original assumption of equilibrium is again invalid and

$$f = \mu_k N = 0.25(52.4) = 13.1 \text{ lb}$$

(d) For impending motion up the plane:

$$f = f_{max} = \mu_s N = 0.30(70.48 - 0.259P)$$

$$\rightarrow f = 21.1 - 0.08P \text{ lb (downward)}$$

$$\Sigma F_x = 0$$

$$f = 0.966P - 25.65$$

Equating the expressions for f shows

$$21.1 - 0.08P = 0.966P - 25.65$$

$$\rightarrow P = 44.7 \text{ lb}$$

Problems

10.1 Force of P = 200 N is acting on a block weighing 600 N. For the case where $\mu_s = 0.30$ and $\mu_k = 0.25$, determine if the block is in equilibrium, sliding up the plane, or sliding down the plane. Also calculate the value of the friction force.
(Ans. Block is sliding down the plane, f = 114.9 N)

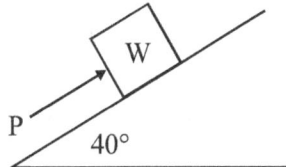

10.2 As shown above, P = 250 N is acting on a block weighing 450 N. For the case where $\mu_s = 0.35$ and $\mu_k = 0.30$, determine if the block is in equilibrium, sliding up the plane, or sliding down the plane. Also calculate the value of the friction force.
(Ans. Block is in equilibrium, f = 39.3 N)

10.3 In the figure below, the weight of the block is 450 lb and the coefficients of static and kinetic friction are 0.32 and 0.26, respectively. If P = 85 lb, determine if the block is in equilibrium and calculate the magnitude and direction of the friction force.
(Ans. The block is in equilibrium, f = 113.1 lb acting up the plane)

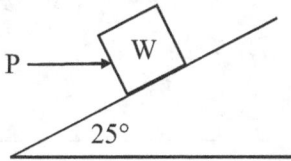

10.4 The weight of the block above is 500 lb and the coefficients of static and kinetic friction are 0.35 and 0.30, respectively. If P = 100 lb, determine if the block is in equilibrium and calculate the magnitude and direction of the friction force.
(Ans. The block is in equilibrium, f = 120.7 lb acting up the plane)

10.5 Consider the block in Problem 10.3 to weigh 350 lb. Compute the coefficient of static friction if a force of P = 75 lb is required to keep the block in equilibrium.
(Ans. $\mu = 0.23$)

10.6 Blocks A, B, and C below have masses of 40 kg, 35 kg, and 30 kg, respectively. The coefficients of static friction are 0.35 between blocks A and B, 0.30 between blocks B and C, and 0.25 between block C and the plane. As θ is slowly increased from 0°, determine which block will begin to slide first and the corresponding value of θ. (Ans. Block C will slide first at θ = 14.0°)

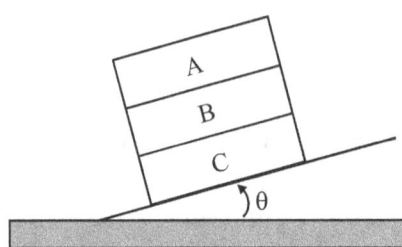

10.7 How will the results of Problem 10.6 change if the coefficients of static friction are 0.30 between blocks A and B, 0.35 between blocks B and C, and 0.40 between block C and the surface.
(Ans. Block A will slide first at θ = 16.7°)

10.8 In Problem 10.3, consider the block to weigh 450 lb. Calculate the coefficient of static friction if a force of P = 120 lb is required to keep the block in equilibrium.
(Ans. μ = 0.18)

10.9 Two blocks on an inclined plane are connected by a rope as shown below. Block B weighs 40 lb and has a coefficient of static friction with the plane of 0.40. Block A has a coefficient of static friction with the plane of 0.20. Find the minimum weight of block A so that motion of the two blocks is impending.
(Ans. W = 8.8 lb)

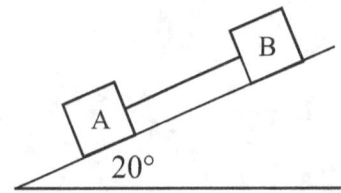

10.10 In the figure above, two blocks on an inclined plane are connected by a rope. Block B weighs 30 lb and has a coefficient of static friction with the plane of 0.50. Block A has a coefficient of static friction with the plane of 0.25. Determine the minimum weight of block A so that motion of the two blocks is impending and determine the tension in the rope.
(Ans. W = 35.8 lb, T = 3.8 lb)

10.11 Blocks A, B, and C have masses of 20 kg, 25 kg, and 30 kg, respectively. The coefficients of static friction are 0.40 between A and B, 0.30 between B and C, and 0.20 between C and the floor. Calculate the smallest value of P that will cause one of the blocks to move and determine which block will move.
(Ans. P = 210.9 N, block B will move)

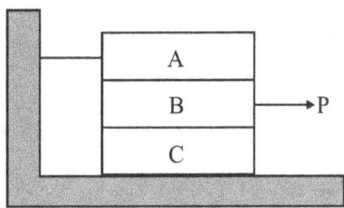

10.12 As shown above, blocks A, B, and C have masses of 35 kg, 30 kg, and 25 kg, respectively. The coefficients of static friction are 0.38 between A and B, 0.30 between B and C, and 0.22 between C and the floor. Determine the smallest value of P that will cause one block to move and determine which block will move.
(Ans. P = 321.8 N, block B will move)

10.13 In the figure below, calculate the smallest value of P that will cause block B to move. All blocks have a mass of 30 kg and the coefficient of static friction is 0.3 between the blocks. Assume the bottom surface is smooth.
(Ans. P = 264.9 N)

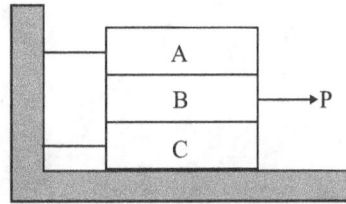

10.14 In Problem 10.6, the masses of blocks A, B, and C are 30 kg, 25 kg, and 20 kg, respectively. The coefficients of static friction are 0.40 between blocks A and B, 0.35 between blocks B and C, and 0.30 between block C and the floor. At what angle will block A begin to slide from atop of block B?
(Ans. θ = 21.8°)

10.15 In Problem 10.13, find the tension in the cable attaching block C to the wall when motion of block B is impending.
(Ans. T_C = 176.6 N)

(This page was intentionally left blank.)

MODULE 11: Centroids

A centroid is defined as the geometrical center of a body. If the body is homogeneous, i.e., the same density throughout, the centroid coincides with the center-of-gravity of the body. If the body is not homogeneous, these points do not coincide. Centroids are used to determine the center-of-gravity of a body and the point within a distributed load where the total load can be considered acting.

Consider a thin flat plate having uniform thickness throughout to be a two-dimensional body. The centroid of such a body can be defined as the geometrical center of the two-dimensional area that defines the shape of this body. The location of the centroid, denoted by the coordinates x^* and y^*, is calculated using the Principle of First Moments of Area as defined by the integral expressions

$$x^* = \frac{\int (x\, dA)}{\int dA}$$

$$y^* = \frac{\int (y\, dA)}{\int dA}$$

Here A is the total area of the shape being considered and, $A = \int d\,A$. The quantities $\int (x\, dA)$ and $\int (y\, dA)$ are called the 'first moments of area'.

For common shapes, simple formulas for x^* and y^* can be used without the need for integration. Table 11.1 summarizes of common shapes and their centroids.

<u>Method of Composite Areas</u>

When a body can be easily divided into several parts whose centroids are known or easily determined, the Method of Composite Areas can determine the centroid of the entire body. This is done by using the following formulas, along with the formulas presented in Table 11.1, for the computation of the coordinates of the centroid of the composite area, x_C^* and y_C^*.

$$x_C^* = \frac{\Sigma(x^* A)}{\Sigma A}$$

$$y_C^* = \frac{\Sigma(y^* A)}{\Sigma A}$$

where the terms $(x^* A)$ and $(y^* A)$ represent the product of the individual centroid components and areas. This method is demonstrated in the examples that follow Table 11.1.

Table 11.1 Centroids of Common Shapes

Rectangular Area:

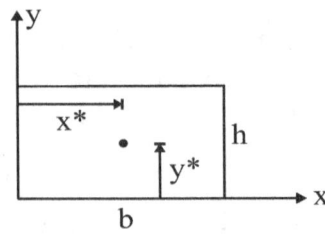

$$x^* = \frac{b}{2} \qquad\qquad y^* = \frac{h}{2}$$

Triangular Area:

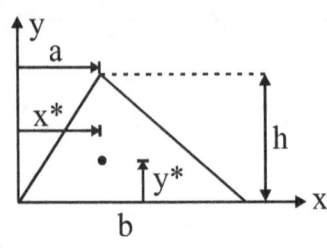

$$x^* = \frac{a+b}{3} \qquad\qquad y^* = \frac{h}{3}$$

Circular Area:

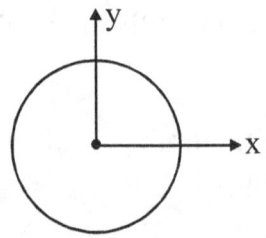

$$x^* = 0 \qquad\qquad y^* = 0$$

Semi-Circular Area:

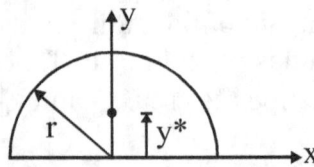

$$x^* = 0 \qquad\qquad y^* = \frac{4r}{3\pi}$$

Quarter-Circular Area:

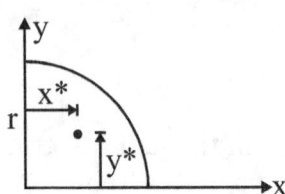

$$y^* = \frac{4r}{3\pi} \qquad\qquad y^* = \frac{4r}{3\pi}$$

Example 11.1

Locate the centroid of the body shown below.

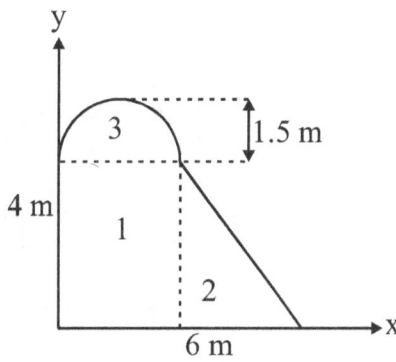

Solution:

This body can easily be broken into three common shapes; a rectangular area, a triangular area, and a semicircular area. These areas are designated as areas 1, 2, and 3, respectively. The table below provides the quantities x^*, y^*, A, x^*A, and y^*A for each area, where x^* and y^* are calculated using the formulas provided in Table 11.1. Note that these values are adjusted to provide the coordinates of the centroid of each area in the coordinate system specified for the entire composite area. The columns for A, x^*A, and y^*A are then added and the results are used to compute x_C^* and y_C^* from the formulas provided above.

Area	A (m²)	x* (m)	y* (m)	x*A (m³)	y*A (m³)
1	12.00	1.50	2.00	18.00	24.00
2	6.00	4.00	1.33	24.00	8.00
3	3.53	1.50	4.64	5.30	16.38
Totals:	21.53			47.30	48.38

For the composite body

$$x_C^* = \frac{\Sigma(x^*A)}{\Sigma A} = \frac{47.30}{21.53} = 2.20 \text{ m}$$

$$y_C^* = \frac{\Sigma(y^*A)}{\Sigma A} = \frac{48.38}{21.53} = 2.25 \text{ m}$$

Example 11.2

The body below contains a circular cutout having a radius of 1.0 in. Compute the centroid of the composite body shown.

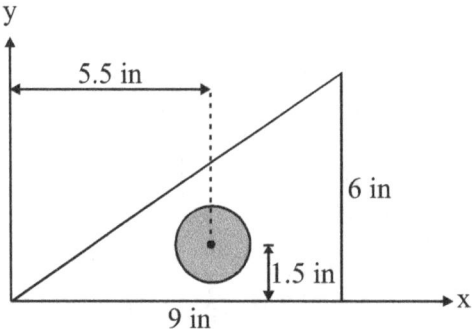

Solution:

The Method of Composite Areas can also be used for a problem containing cutouts. However, the cutout sections must be considered having a negative area when computing the tabular quantities used in the solution. Considering the solid triangular area to be area 1 and the circular cutout to be area 2, the following table of values can be constructed using the formulas for common shapes.

Area	A (in²)	x* (in)	y* (in)	x*A (in³)	y*A (in³)
1	27.00	6.00	2.00	162.00	54.00
2	−3.14	5.50	1.50	−17.27	−4.71
Totals:	23.86			144.73	49.29

For the composite body

$$x_C^* = \frac{\Sigma(x^*A)}{\Sigma A} = \frac{144.73}{23.86} = 6.07 \text{ in}$$

$$y_C^* = \frac{\Sigma(y^*A)}{\Sigma A} = \frac{49.29}{23.86} = 2.07 \text{ in}$$

Problems

11.1 Compute the x-centroid of the area below if a = 3.5 ft.
(Ans. $x_C^* = 1.92$ ft)

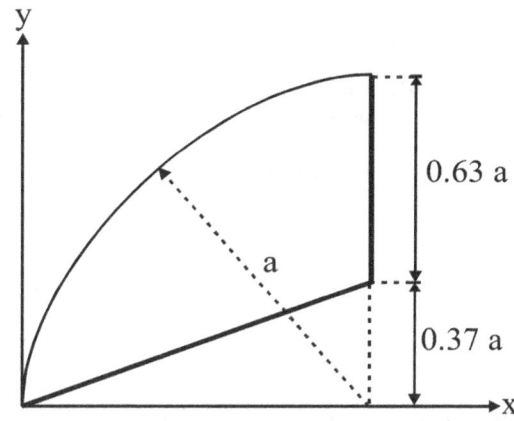

11.2 Find the y-centroid of the area above if a = 2.0 ft.
(Ans. $y_C^* = 1.04$ ft)

11.3 For the figure in Problem 11.1, calculate the x-centroid in terms of the unknown dimension a.
(Ans. $x_C^* = 0.55a$ in)

11.4 Compute the x-centroid of the area below. The radius of the cutout is 3.9 m.
(Ans. $x_C^* = 10.95$ m)

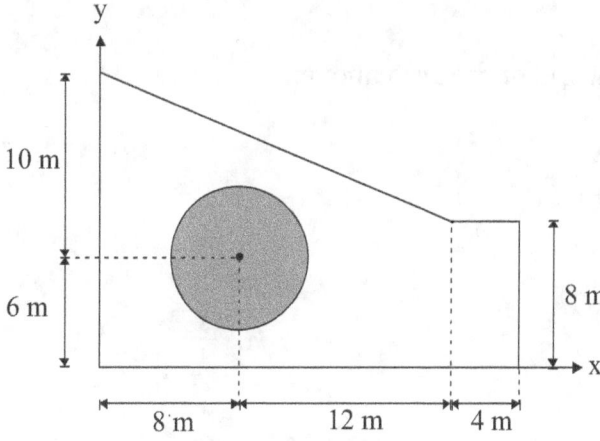

11.5 Calculate the y-centroid of the area above.
(Ans. $y_C^* = 5.95$ m)

11.6 Determine the x-coordinate of the centroid for the triangular area and cutout below. The units of H are meters.
(Ans. $x_C^* = 0.82H$ m)

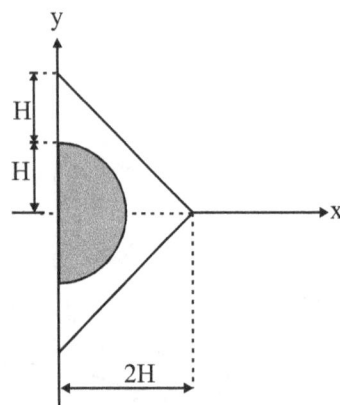

11.7 Compute the y-centroid of the area shown for the case of L = 7.5 in.
(Ans. $y_C^* = 6.57$ in)

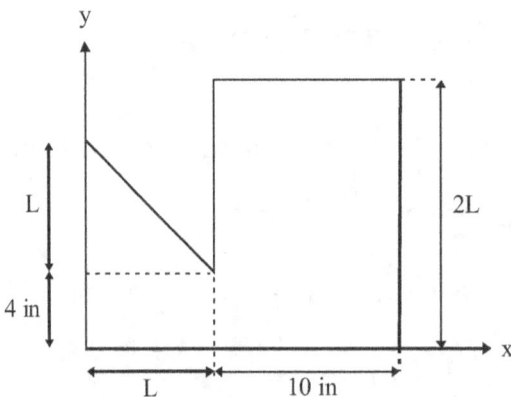

11.8 Calculate the x-centroid of the area above.
(Ans. $x_C^* = 9.89$ in)

11.9 Find the centroid of the area below with respect to the given coordinate system.
(Ans. $x^*_C = 0.010$ m, $y^*_C = 0.013$ m)

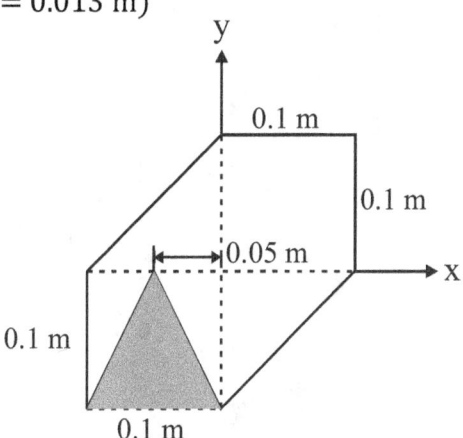

11.10 In Problem 11.1, calculate the y-centroid in terms of the unknown dimension a.
(Ans. $y^*_C = 0.52a$ in)

11.11 The plate shown below includes a set of 2.0 in diameter cutouts. Compute the x-coordinate of the centroid of this plate.
(Ans. $x^*_C = 7.80$ in)

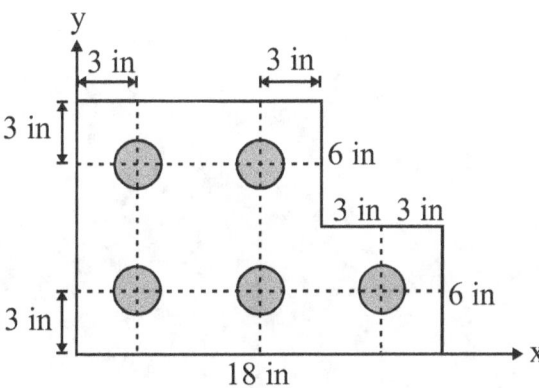

11.12 Find the y-coordinate of the centroid of the plate shown above.
(Ans. $y^*_C = 5.40$ in)

1113 Determine the centroid of the body in Example 11.2 in terms of the radius, r, of the circular cutout.
(Ans. $x^*_C = \dfrac{5.5(9.376 - r^2)}{8.594 - r^2}$ in, $y^*_C = \dfrac{1.5(11.459 - r^2)}{8.594 - r^2}$ in)

11.14 Solve Example 11.2 for the case of the cutout being located at the coordinates (7.0, 2.5) in the given coordinate system.
(Ans. $x^*_C = 5.87$ in, $y^*_C = 1.93$ in)

11.15 Calculate the centroid of the area below with respect to the coordinate system provided.
(Ans. $x_C^* = 0.11$ m, $y_C^* = 0.15$ m)

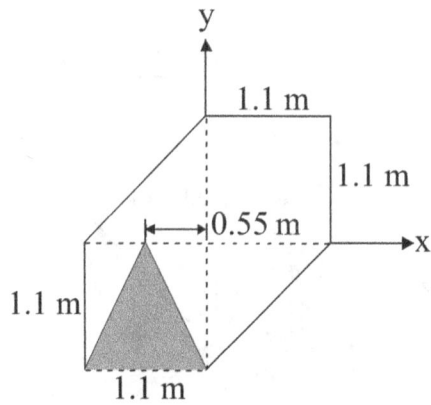

MODULE 12: Loads on Beams

Beams

Beams are structural members that provide a resistance to bending from applied loads. They are long prismatic bars and loads are most often applied normal to the longitudinal axis of the bar. These applied loads can be concentrated loads, various types of distributed loads, or concentrated couples. Methods of analyzing beams for various types of loading conditions are presented in this and the next module.

Prior to the study of distributed loadings on beams, however, it is important to understand the two types of beams. Beams that are supported in a manner that the external support reactions can be calculated by the methods addressed in statics are called 'statically determinate' beams. Beams that are supported by more external supports than are necessary for equilibrium are called 'statically indeterminate' beams. More advanced analysis methods are required to determine the support reactions in beams that are statically indeterminate. Figure 12.1 gives examples of both statically determinate and statically indeterminate beams. Only statically determinate beams are addressed in this text.

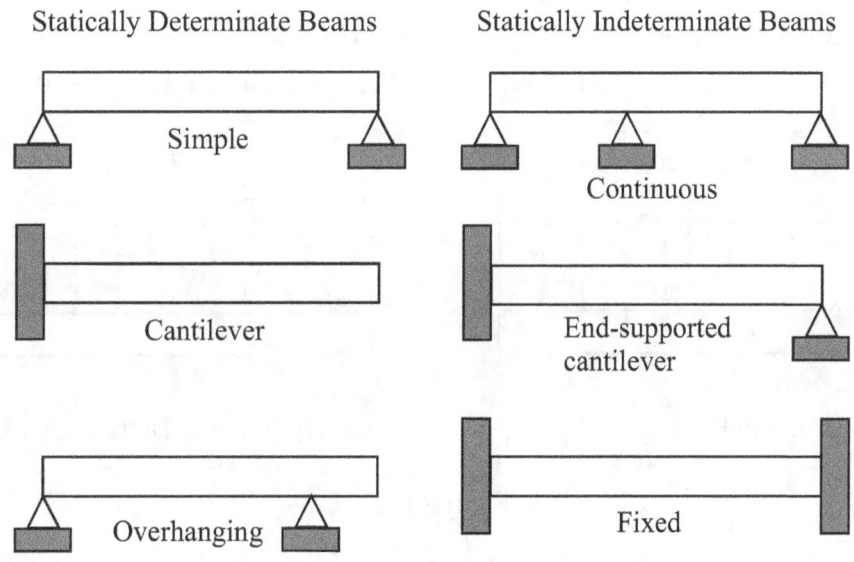

Figure 12.1

Types of Loading

Distributed loads are loads that are applied over a portion of the beam's length with a specified intensity. This intensity, w, is expressed as a force per unit length of beam and can be uniform or variable and continuous or discontinuous. Loadings that are uniformly distributed, i.e., constant, or distributed as a linear function of the beam's length, are easily handled. Examples of uniformly and linearly varying distributed loads are shown in Figure 12.2.

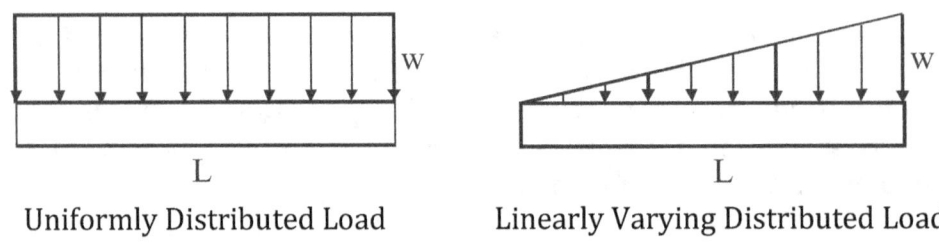

Uniformly Distributed Load   Linearly Varying Distributed Load

Figure 12.2

The effects of distributed loads are analyzed by replacing the distributed load with a concentrated load having a magnitude equal to the total load, i.e., area under the curve of w, and placed at the centroid of the area representing the load distribution. For a uniform load, the concentrated load would be equal to P = wL and placed at the center of the load's span. For a linearly distributed load, the concentrated load would be equal to P = 0.5 wL and placed at a distance of $\frac{L}{3}$ from the non-zero side of the load distribution. Figure 12.3 shows how distributed loads can be replaced with concentrated loads at their centroids.

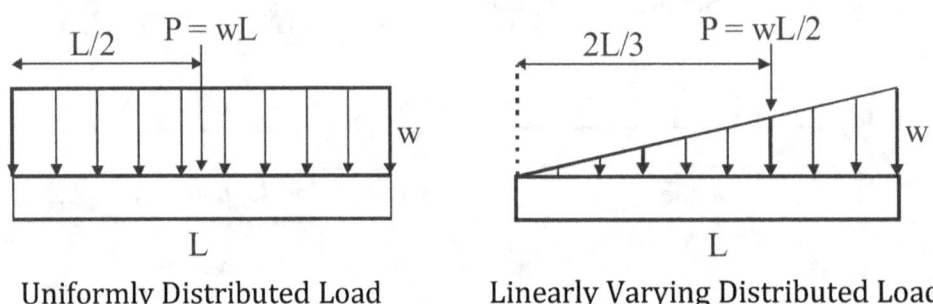

Uniformly Distributed Load   Linearly Varying Distributed Load

Figure 12.3

Distributed loads can also comprise a combination of a uniformly and a linearly varying distribution. An example of this condition is shown in Figure 12.4.

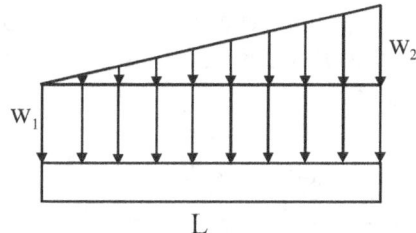

Figure 12.4

Here, the distributed loads can be replaced with concentrated loads by one of two methods. One method is to place a concentrated load for each distributed load at the centroid of each area, as was done in Figure 12.3. Another method is to place a single concentrated load at the centroid of the combined areas representing the uniform load distribution and the linearly varying load distribution. In this case, the centroid for the combined loads must first be determined by the Method of Composite Areas and will therefore require additional calculations.

Example 12.1

For the simple beam shown below, calculate the reactions at the supports.

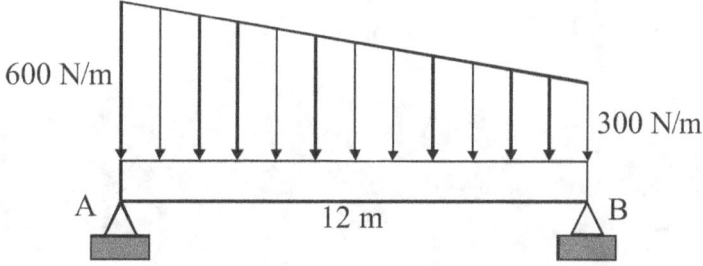

Solution:

The distributed loads for this beam comprise a uniform load of $w_1 = 300 \frac{N}{m}$, and a linearly varying load $w_2 = 300 \frac{N}{m}$ at the left support and decreasing to zero at the right support. These distributed loads can be replaced with two concentrated loads

where

$P_1 = w_1 L = 300(12.0) = 3,600 \text{ N}$

$P_2 = w_2 \left(\frac{L}{2}\right) = 300 \left(\frac{12.0}{2}\right) = 1,800 \text{ N}$

These concentrated loads are placed at the centroids of the rectangular area (for the constant load) and the triangular area (for the linearly varying load) as shown on the free-body diagram below.

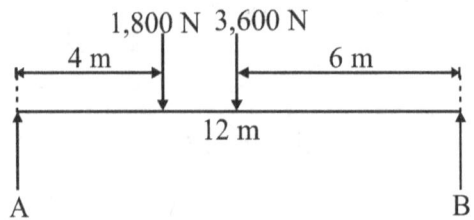

$\sum F_y = 0$

$A + B = 5,400$

$\rightarrow A = 5,400 - B$

$\sum M_A = 0$

$4\,\mathbf{i} \times (-1,800\,\mathbf{j}) + 6\,\mathbf{i} \times (-3,600\,\mathbf{j}) + 12\,\mathbf{i} \times B\,\mathbf{j} = 0$

$-7,200\,\mathbf{k} - 21,600\,\mathbf{k} + 12B\,\mathbf{k} = 0$

$12B = 28,800$

$\rightarrow B = 2,400\text{ N}$

$\rightarrow A = 3,000\text{ N}$

Example 12.2

For the cantilever beam, determine the reactions at the wall.

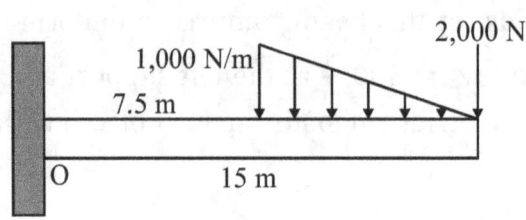

Solution:

The linearly varying distributed load can be replaced by a concentrated load of

$$P = 1{,}000 \left(\frac{7.5}{2}\right) = 3{,}750 \text{ N}$$

P is at the centroid of the triangular area, which is a 5 m from the free end of the beam. The free-body diagram can then be drawn as

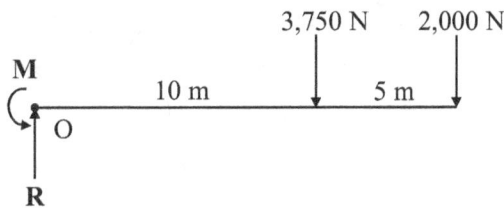

$$\sum F_y = 0$$

$$R - 3{,}750 - 2{,}000 = 0$$

$$\rightarrow R = 5{,}750 \text{ N}$$

$$\sum M_O = 0$$

$$10\,\mathbf{i} \times (-3{,}750\,\mathbf{j}) + 15\,\mathbf{i} \times (-2{,}000\,\mathbf{j}) + \mathbf{M} = 0$$

$$-37{,}500\,\mathbf{k} - 30{,}000\,\mathbf{k} + M\,\mathbf{k} = \mathbf{0}$$

$$\rightarrow M = 67{,}500 \text{ N-m}$$

$$\rightarrow \mathbf{M} = 67{,}500\,\mathbf{k} \text{ N-m}$$

Problems

12.1 Determine the reactions at the supports for the beam shown and the case of
$w = 750\,\frac{lb}{ft}$.
(Ans. A = 5,875 lb, B = 3,750 lb)

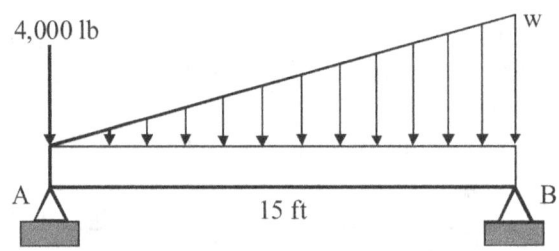

12.2 Calculate the reaction forces at the supports for the loading conditions shown below.
(Ans. A = 617 N, B = 758 N)

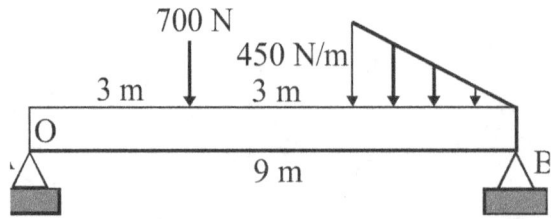

12.3 Compute the reactions at the supports for the loading conditions shown.
(Ans. A. = 2,925 N, B = 675 N)

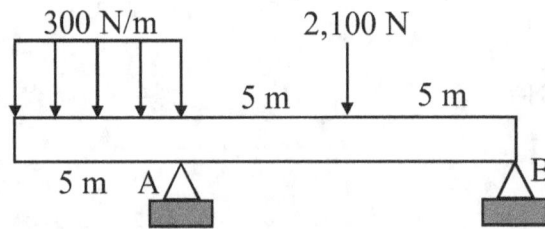

12.4 Find the reaction forces at the supports for the simple beam shown below if
$w = 850 \frac{\text{lb}}{\text{ft}}$.
(Ans. A = 3,271 lb, B = 2,504 lb)

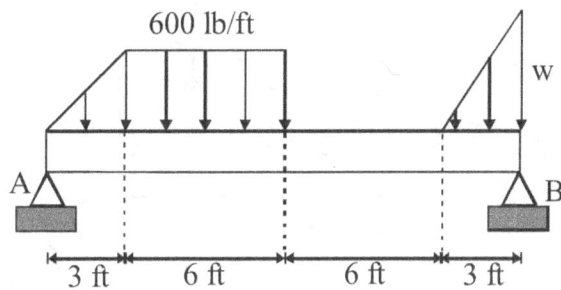

12.5 In the figure above, calculate the reaction forces at the supports in terms of an unknown distributed load w for the simple beam shown.
(Ans. A = 3,200 + 0.08w lb, B = 1,300 + 1.42w lb)

12.6 Determine the reactions at the wall of the cantilever beam shown. The intensity of the distributed load at the wall is $410 \frac{\text{lb}}{\text{ft}}$ and L = 15 ft.
(Ans. R = 10,200 lb, M = 86,625 lb-ft)

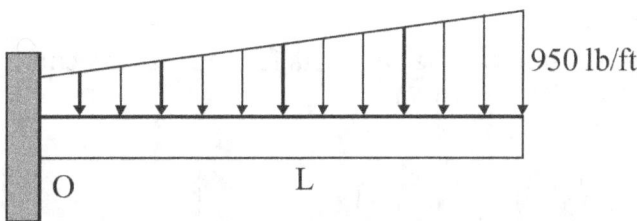

12.7 If L = 14 m, find the distance d, so that the reaction forces at the supports are equal.
(Ans. d = 4.4 m)

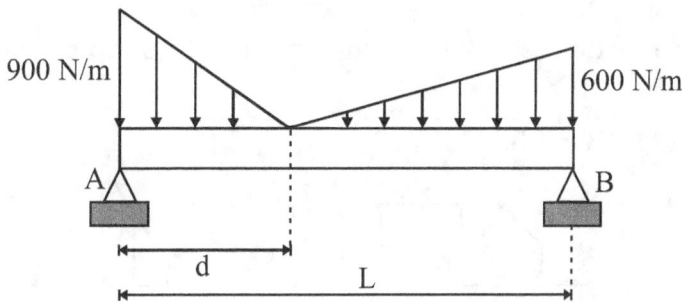

12.8 In the figure above, calculate d so that A carries 60% of the applied load.
(Ans. d = 8.7 m)

12.9 Determine the reactions at the supports for the beam shown and the case of
$w = 600 \frac{lb}{ft}$.
(Ans. A = 4,401 lb, B = 2,799 lb)

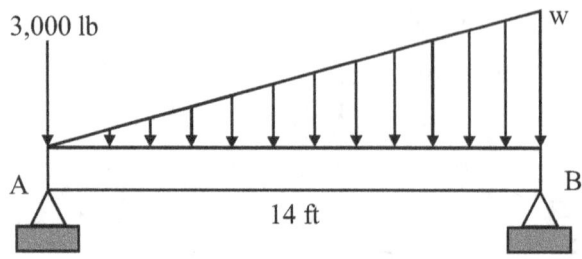

12.10 Compute the reactions at the supports for the loading conditions shown.
(Ans. A = 300 N, B = 650 N)

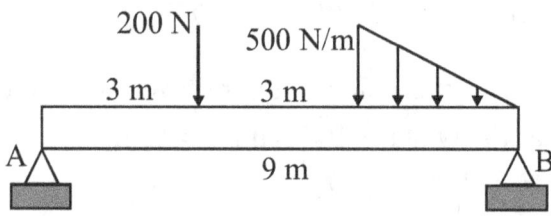

12.11 Calculate the reactions at the supports for the loading conditions shown.
(Ans. A. = 2,150 N, B = 650 N)

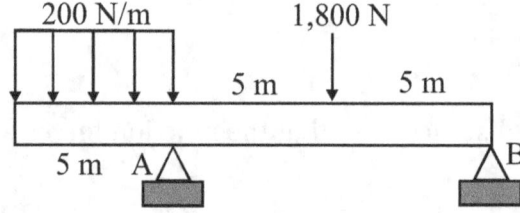

12.12 Determine the reaction forces at the supports for the simple beam shown below if
$w = 700 \frac{lb}{ft}$.
(Ans. A = 1,817 lb, B = 1,383 lb)

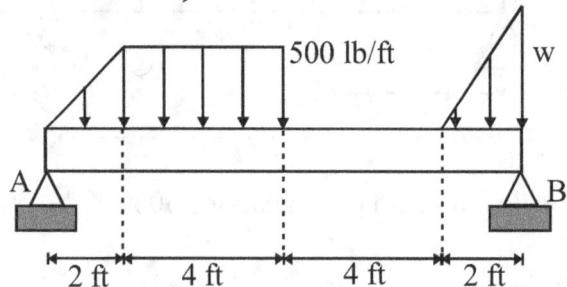

12.13 Solve Problem 12.12 in terms of the unknown distributed load w.
(Ans. A = 1,778 – 0.06w lb, B = 722 + 0.94w lb)

12.14 If L = 10 m, find the distance d so that the reaction forces at the supports are equal.
(Ans. d = 3.2 m)

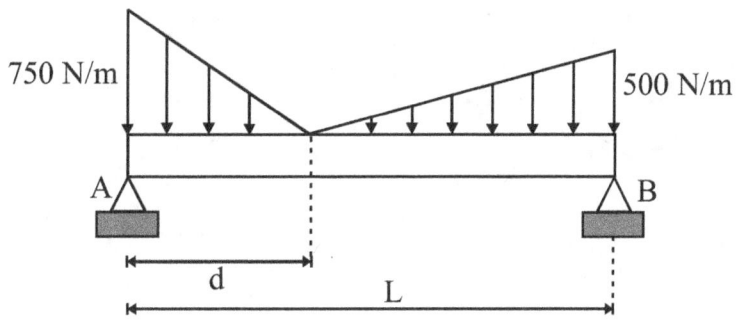

12.15 Solve Problem 12.14 for the case where support A carries 60% of the total applied load.
(Ans. d = 6.2 m)

(This page was intentionally left blank.)

MODULE 13: Nonlinear Distributed Loads on Beams

In addition to uniformly and linearly varying distributed loads, it is common for beams to be loaded with distributed loads that are a nonlinear function of the beam's length. This distribution could be sinusoidal, quadratic, cubic, parabolic or some combination of these functions, or it could be a more complicated function of the distance along the beam on which the loading is applied. Whatever the case, the loading distribution is given, or easily determined, as a function a distance x along the beam. Here, the same general procedure is used to analyze the loading effects on the beam as was described in earlier. First, the total load provided by the distributed load will be determined, and second, the centroid of the area representing the loading pattern will be located so the total load can be applied at this point. The difference between these types of loads and loads having a uniformly or linearly varying distribution, however, is that it is slightly more difficult to determine the centroid and total load in nonlinear loading patterns. In the general case, a distributed load has the form of w = w(x), where x is the distance measured from the starting or initial point, e.g., usually the left side, of the applied load. The total load represented by this loading pattern over the application span can be calculated by the integral expression

$$P = \int (w \, dx)$$

Since w is most often a simple function of x, this integral is easy to evaluate over the length of the loading span. Once this total load is obtained, the centroid of the area representing the loading pattern can be found using the integral expression for the first moment of the load given by

$$x^* = \frac{\int (xw \, dx)}{P}$$

Finding the x-coordinate of the centroid is also easy to evaluate over the loading span. However, the functional form of w may involve constants that must be evaluated using the load values at the initial point and/or final point of the loading pattern. Once both the load and the centroid are determined, the beam support reactions are evaluated using the standard methods previously discussed. Of course, nonlinear distributed loads can also be applied in combination with uniformly, linearly varying, or concentrated loads.

Example 13.1

For the simple beam shown below, the loading distribution is given by the function, $w = w_0 + kx^2$. Determine the reactions at the supports under this loading condition.

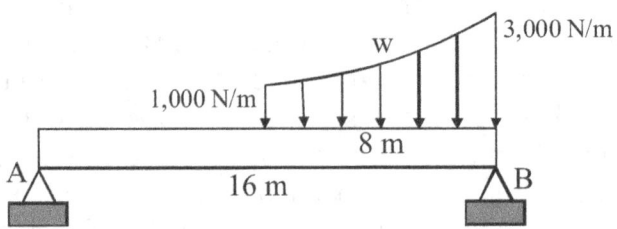

Solution:

The first step in the analysis of the given loading condition is to determine the constants w_0 and k in the loading function, w. This is done by using the values of the load at both the initial point of load application ($x = 0$) and the final point of load application ($x = 8$ m). The initial and final values of the load are

at $x = 0$ m: $w = 1{,}000 \, \frac{N}{m}$

at $x = 8$ m: $w = 3{,}000 \, \frac{N}{m}$

Using this information to evaluate w_0 and k shows

$$1{,}000 = w_0 + k(0)^2$$

$$\rightarrow w_0 = 1{,}000 \, \frac{N}{m}$$

$$3{,}000 = 1{,}000 + k(8)^2$$

$$\rightarrow k = 31.25 \, \frac{N}{m^3}$$

The loading function can then be expressed as

$$w = 1{,}000 + 31.25x^2 \, \frac{N}{m}$$

The total load provided by w over the span of 8 m is

$$P = \int (w \, dx) = \int [1{,}000 + 31.25x^2] \, dx \text{ N}$$

$$P = \left[1{,}000x + 31.25 \left(\frac{x^3}{3}\right)\right] \text{ evaluated from } x = 0 \text{ to } x = 8 \text{ m}$$

$$\rightarrow P = 1{,}000(8) + 31.25 \left[\frac{(8)^3}{3}\right] = 13{,}333.3 \text{ N}$$

Calculation of the centroid location shows

$$x^* = \frac{\int (xw \, dx)}{P}$$

$$x^* = \frac{\int x(1{,}000 + 31.25x^2) \, dx}{P}$$

$$x^* = \frac{\int (1{,}000x + 31.25x^3) \, dx}{P}$$

$$x^* = \frac{500x^2 + \left(\frac{31.25}{4}\right)x^4}{13{,}333.3} \text{ evaluated from } x = 0 \text{ to } x = 8 \text{ m}$$

$$x^* = \frac{500(8)^2 + \left(\frac{31.25}{4}\right)(8)^4}{13{,}333.3}$$

$$\rightarrow x^* = 4.80 \text{ m}$$

This result indicates that the centroid of the loaded area will be 4.8 m to the right of the start of the loading pattern, or a total distance of 12.8 m from point A on the beam. The free-body diagram is

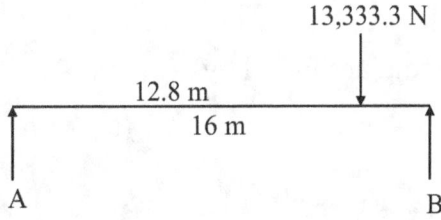

Solving for the reaction forces at A and B by summing the y forces and summing the moments about point A or point B, give the solution as

A = 2,666.7 N

B = 10,666.7 N

Example 13.2

Determine the reactions at the wall for the beam shown below. The applied load is given as $w = kx^{1/2}$.

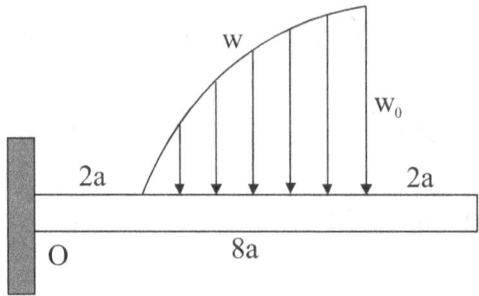

Solution:

The end conditions of the load are

at $x = 0$: $w = 0$

at $x = 4a$: $w = w_0$

Using this information shows

$$w = w_0 = k(4a)^{1/2}$$

$$\rightarrow w_0 = 2ka^{1/2}$$

$$\rightarrow k = \frac{w_0}{2a^{1/2}}$$

$$\rightarrow w = \frac{w_0}{2}\left(\frac{x}{a}\right)^{1/2}$$

The total load and centroid are found by

$$P = \int \left(\frac{w_0}{2a^{1/2}}\right) x^{1/2} \, dx$$

$$P = \left(\frac{w_0}{3a^{1/2}}\right) x^{3/2} \quad \text{evaluated from } x = 0 \text{ to } x = 4a$$

$$\rightarrow P = 2.67aw_0$$

$$x^* = \frac{\int xw \, dx}{P}$$

$$x^* = \frac{\int x\left(\frac{w_0}{2a^{1/2}}\right)x^{1/2}\,dx}{P}$$

$$x^* = \frac{\int \left(\frac{w_0}{2a^{1/2}}\right)x^{3/2}\,dx}{P}$$

$$x^* = \left(\frac{1}{2.67aw_0}\right)\left(\frac{w_0}{5a^{1/2}}\right)x^{5/2} \quad \text{evaluated from } x = 0 \text{ to } x = 4a$$

$$\rightarrow x^* = 2.4a$$

Therefore, a concentrated load equal to P can be applied at a point a distance of 2.4a to the right of the left side of the distributed load, or 4.4a from the wall. The   free-body diagram then becomes

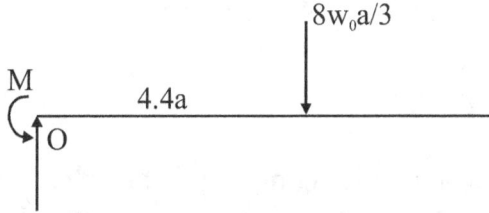

Solving the equilibrium equations gives the reactions at the wall as

$$O = 2.67\,w_0a$$

$$M = 11.73\,w_0a^2$$

Problems

13.1 Determine the reactions at supports A and B below for the loading pattern of
$w = 12x^2 \frac{N}{m}$.
(Ans. A = 512 N, B = 1,536 N)

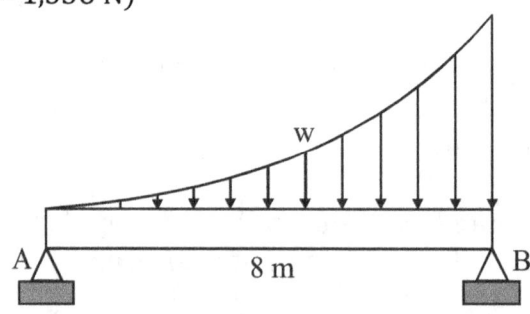

13.2 Calculate the reactions at supports A and B above for the loading pattern of
$w = 10x^3 \frac{N}{m}$.
(Ans. A = 2,048 N, B = 8,192 N)

13.3 The figure below shows an overhanging simple beam supporting a distributed load
given by $w = x + kx^2$. The value of this load at the right side of the distribution is
$650 \frac{lb}{ft}$. Find the reactions at the supports.
(Ans. A = 1,415 lb, B = 2,095 lb)

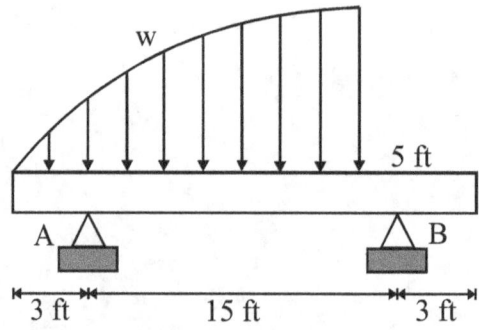

13.4 For the beam shown, the distributed load is given as $w = 35 \sin\left(\frac{\pi x}{L}\right)$ lb. Compute the reaction force and moment at the wall of the beam in terms of L (ft).
(Ans. R = 22.3L lb, M = 11.1L^2 ft-lb)

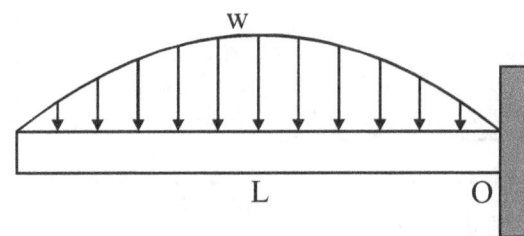

13.5 If the distributed load on the beam above is given as $w = 20 \sin\left(\frac{Bx}{L}\right)$ lb, determine the reactions at the wall in terms of L ft.
(Ans. R = 12.7L lb, M = 6.4L^2 lb-ft)

13.6 In Example 13.2, find the reaction forces at the supports if the beam has simple supports at each end.
(Ans. A = 1.2w_oa, B = 1.5w_oa)

13.7 The beam in the figure below is subjected to a distributed load given by $w = 250x^{1/2} \frac{N}{m}$. Compute the reactions at the supports for this loading condition.
(Ans. A = 4,472 N, B = 10,435 N)

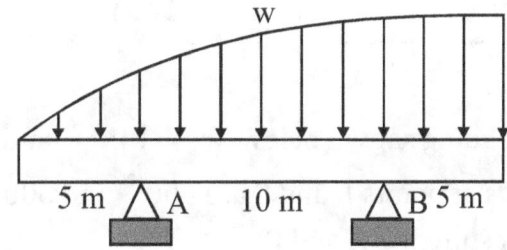

13.8 Calculate the reactions at the supports of the beam above for the distributed load given by $w = 300x^{1/3} \frac{N}{m}$.
(Ans. A = 4,361 N, B = 7,854 N)

13.9 Determine the reactions at supports A and B below for the loading pattern of $w = 10x^2 \frac{N}{m}$.
(Ans. A = 104 N, B = 313 N)

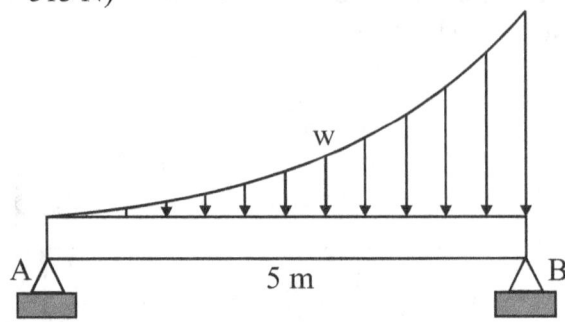

13.10 The figure below shows an overhanging simple beam supporting a distributed load given by $w = x + kx^2 \frac{lb}{ft}$. The value of this load at the right side of the distribution is $500 \frac{lb}{ft}$.
Find the reactions at the supports.
(Ans. A = 897 lb, B = 1,469 lb)

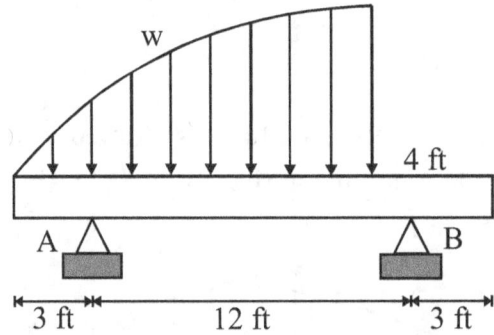

13.11 For the distributed loading shown below, $w_1 = k_1 x^2 \frac{N}{m}$ and $w_2 = k_2 + k_3 x^2 \frac{N}{m}$, and the values of the loads at points B and C are $400 \frac{N}{m}$ and $500 \frac{N}{m}$, respectively. Calculate the reaction forces at supports A and D.
(Ans. A = 1,181 N, D = 1,653 N)

13.12 For the distributed loading shown, $w_1 = k_1x^2 \frac{N}{m}$ and $w_2 = k_2 + k_3x^2 \frac{N}{m}$, and the values of the loads at points B and C are $200 \frac{N}{m}$ and $300 \frac{N}{m}$, respectively. Compute the reaction forces at supports A and D.
(Ans. A = 482 N, D = 744 N)

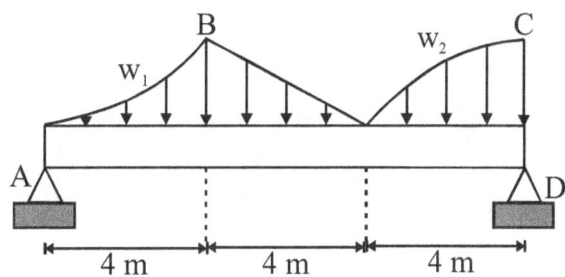

13.13 The beam in the figure below is subjected to a distributed load given by $w = 300x^{3/2} \frac{N}{m}$. Determine the reactions at the supports for this loading condition.
(Ans. A = 3,840 N, B = 8,960 N)

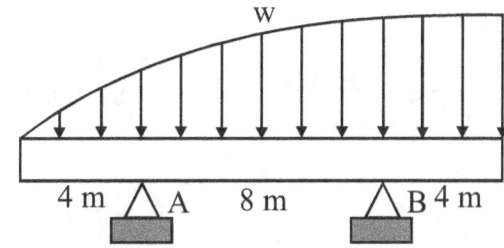

13.14 Solve Problem 13.9 for the case where $w = 10x^3 \frac{N}{m}$.
(Ans. A = 313 N, B = 1,250 N)

13.15 Calculate the reactions at the supports in Problem 13.13 for the distributed load given by $w = 300x^{1/3} \frac{N}{m}$.
(Ans. A = 3,240 N, B = 5,832 N)

(This page was intentionally left blank.)

INDEX

(This page was intentionally left blank.)

www.ingramcontent.com/pod-product-compliance
Lightning Source LLC
Chambersburg PA
CBHW081106290526
45795CB00006B/2021